# Convergence Analysis of Recurrent Neural Networks

# Network Theory and Applications

Volume 13

# CONVERGENCE ANALYSIS OF RECURRENT NEURAL NETWORKS

ZHANG YI
School of Computer Science and Engineering
University of Electronic Science and Technology of China
Chengdu 610054
People's Republic of China

K. K. TAN
Department of Electrical and Computer Engineering
The National University of Singapore
4 Engineering Drive 3
117576 Singapore

**Kluwer Academic Publishers**
Boston/Dordrecht/London

Distributors for North, Central and South America:
Kluwer Academic Publishers
101 Philip Drive
Assinippi Park
Norwell, Massachusetts 02061 USA
Telephone (781) 871-6600
Fax (781) 871-6528
E-Mail <kluwer@wkap.com>

Distributors for all other countries:
Kluwer Academic Publishers Group
Post Office Box 322
3300 AH Dordrecht, THE NETHERLANDS
Telephone 31 78 6576 000
Fax 31 78 6576 474
E-Mail <orderdept@wkap.nl>

Electronic Services <http://www.wkap.nl>

**Library of Congress Cataloging-in-Publication**

Yi, Zhang/ Tan, K.K.
Convergence Analysis of Recurrent Neural Networks
ISBN 1-4020-7694-0


*Printed on acid-free paper.*

Printed in United Kingdom by Biddles/IBT Global

# Contents

# List of Figures

# Preface

Since the outstanding and pioneering research work of Hopfield on recurrent neural networks (RNNs) in the early 80s of the last century, neural networks have rekindled strong interests in scientists and researchers. Recent years have recorded a remarkable advance in research and development work on RNNs, both in theoretical research as well as actual applications. The field of RNNs is now transforming into a complete and independent subject. From theory to application, from software to hardware, new and exciting results are emerging day after day, reflecting the keen interest RNNs have instilled in everyone, from researchers to practitioners.

RNNs contain feedback connections among the neurons, a phenomenon which has led rather naturally to RNNs being regarded as dynamical systems. RNNs can be described by continuous time differential systems, discrete time systems, or functional differential systems, and more generally, in terms of non-linear systems. Thus, RNNs have to their disposal, a huge set of mathematical tools relating to dynamical system theory which has turned out to be very useful in enabling a rigorous analysis of RNNs.

RNNs have been found many applications, including applications relating to associative memory, image processing, pattern recognisation, etc. Their dynamical properties play a crucial and important role to enable these successful practical applications of RNNs. From the engineering point of view, neural network models, which possess simple structures and well understood dynamic properties are of particular interest to consider in a practical application. It is essential and necessary to understand these dynamical properties in order to be able to choose and design effective control parameters.

Among the dynamical properties, convergence is one of the most important issues of concern in a practical application. In fact, the viability of many applications of RNNs strongly depend on their convergence properties. Without a proper understanding of the convergence properties of RNNs, many of these applications would not be possible. Furthermore, convergence conditions are essential for the actual design of RNNs.

Essentially, the study of convergence of RNNs broadly seeks to derive conditions which can guarantee that a RNN will be convergent. Since these conditions will be used in the design of RNNs, simple and relaxed convergent conditions of RNNs, which can be easily verified, are more amenable to practical use. Although many RNNs are described in terms of nonlinear systems, few of the existing results in the stability of nonlinear theory can meet the requirements. In fact, there have arised certain problems where RNNs can pose new challenges to mathematicians.

This book is focused mainly on the study of the convergence of RNNs. The RNN models explored in the book includes Hopfield RNNs, Cellular RNNs, Lotka-Volterra RNNs, RNNs with unsaturating activation functions, and discrete time RNNs, etc. RNNs with time delays will be of particular interest as it is well known that delays are important and commonly occurring parameters in RNNs. In biological neural networks, real neurons have integrative time delays due to capacitive effects. In artificial neural networks, time delays are inherently present due to hardware characteristics such as switching delays, parameter variability, parasitic capacitance, etc. RNNs with delays have found applications in the processing and compression of images, etc. Besides the direct applications, it is also useful to analyse networks with delays since delays can drastically change the dynamics of a RNN. A stable RNN may become unstable when a small time delay is added. The convergence analysis of RNNs with delays aims to develop conditions for convergence which can be robust to the destabilising effects of delays.

Two kinds of convergence of RNNs will be studied in this book: monostability and multistability. Both of these convergence characteristics have direct implications on the applications of RNNs. Three methods will be used in the study: Lyapunov functions method, energy functions method, and inequalities analysis method. The analysis is focused on the nonlinear characteristics of RNNs. Rigorous analysis will be carried out where possible in these two kinds of convergence analysis of RNNs.

The book is organised as follows. Chapter 1 provides a brief introduction to RNNs. Some frequently used notations and terminology are given. Basic discussions on energy functions are presented which should lay the foundation for subsequent chapters on multistability analysis.

In Chapter 2, the convergence of Hopfield RNNs is discussed. Complete convergence, local stability as well as global stability and the estimation of the convergence rate will be discussed.

In Chapter 3, the convergence of Cellular Neural Networks (CNNs) is analysed. The analysis will range from exponential convergence, estimate of exponential convergence rates, complete stability, to global stability of CNNs. Both CNNs, with and without delays, will be considered.

In Chapter 4, the convergence of Lotka-Volterra RNNs with delays will be addressed. Both multistability and monostability are analysed. Conditions based on local inhibition which allow networks to possess a multistability property are obtained. The estimation of attractors will be highlighted. By assuming the delays are in $L^2$ space and the symmetry on synaptic connection matrices, complete stability is proven.

In Chapter 5, a typical interesting model of RNNs is discussed, i.e., RNNs with unsaturating piecewise linear transfer functions. This model has attracted much attention recently for its potential in applications. Some hardware implementations based on this model are now available. The chapter will focus on the study of the multistability of this model. Three basic properties: boundedness, global attractivity, and complete stability are discussed by applying local inhibition on neurons.

In Chapter 6, under the assumption that the activation functions are global Lipschitz functions (GLF), the global stability of RNNs with constant and variable delays are studied. Absolute periodicity and absolute stability of RNNs with delays are also discussed in this chapter.

In Chapter 7, other special types of RNNs are presented and studied. The concept of Input-to-State Convergence (ISC) is introduced to a model of RNNs with variable inputs. Some conditions for ISC are established. Detailed convergence analysis is done on a specific model of RNNs which can be used for extracting eigenvectors and eigenvalues of any symmetric matrix.

Finally, in Chapter 8, discrete time RNNs are addressed. A section is devoted to analyse the multistability of discrete time RNNs with unsaturating piecewise linear transfer functions. A model of discrete time RNNs is also presented to deal with the winner-take-all problem.

The contents of this book are mainly based on some of the research work of the authors in RNNs in recent years. While the main objective is to disseminate these results in a unified and comprehensive manner, it is also hoped that the book can be useful to a wide range of readers by furnishing some basic information, methods and tools commonly used in the analysis and design of RNNs. Undoubtedly, there are other excellent work in this area which we hope to have included in the references for the readers. We should also like to point out that at this time of writing, many problems relating to RNNs have remained unresolved and the book may contain personal views and conjecture of the authors which may not appeal to all sectors of the readers. To this end, we welcome criticisms and suggestions from readers and look forward to discussion and collaboration on the topic.

ZHANG YI, K. K. TAN

# Acknowledgments

This book was supported in part by the National Science Foundation of China under Grant 69871005.

***This book is dedicated to all of our loved ones.***

Chapter 1

# INTRODUCTION

## 1. Introduction

### 1.1 Recurrent Neural Networks

Generally, neural networks can be divided into two large classes. One class contains feedforward neural networks (FNNs), and the other contains recurrent neural networks (RNNs). This book focused on RNNs only. The essential difference between FNNs and RNNs is the presence of a feedback mechanism among the neurons in the latter. A FNN is a network without any feedback connections among its neurons, while a RNN has at least one feedback connection. Since RNNs allow feedback connections in neurons, the network topology can be very general: any neuron can be connected to any other, even to itself. Allowing the presence of feedback connections among neurons has an advantage, it leads naturally to an analysis of the networks as dynamic systems, in which the state of a network, at one moment in time, depends on the state at a previous moment in time. The topology of RNNs is shown in Figure 1.1.

The dynamics of a RNN can be then described by a continuous time differential system

$$\dot{x}(t) = f(x(t)), \quad t \geq 0,$$

or a discrete time system

$$x(k+1) = g(x(k)), \quad k = 1, 2, \cdots$$

where $x$ denotes the state of the corresponding network, $f$ and $g$ are some suitable mappings. If one considers transmission delays among neurons, then a RNN model can be described by functional differential system

$$\dot{x}(t) = F(x_t), \quad x_t = x(t+\theta), \quad \theta \in [-\tau, 0],$$

where $\tau > 0$ is some constant and $F$ is a functional.

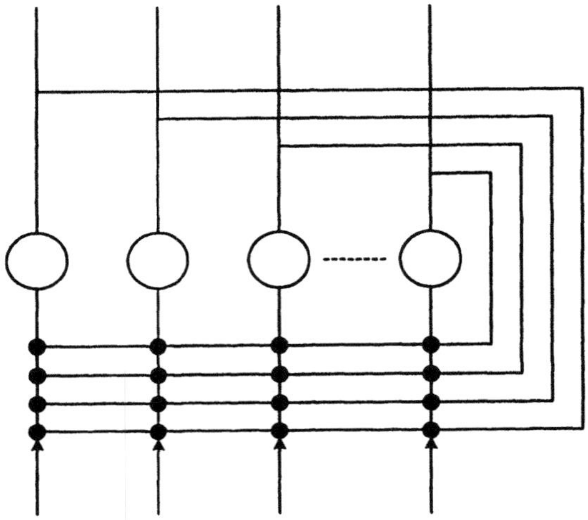

*Figure 1.1.* Topology of RNNs.

Formulating RNNs into differential systems, discrete systems, or functional differential systems provides powerful methods to analyse the properties of RNNs. Some existing mathematical theories of differential systems, discrete systems and functional differential systems can be then applied.

Time delays are important parameters of RNNs. In fact, delays exist both in biological neural networks and artificial neural networks. In hardware implementation of RNNs, time delays occur due to the finite switching speed of the amplifiers. Some delayed RNNs have direct applications, for example, in processing of moving images, image compression, etc. In addition, delays may affect the dynamics of RNNs. A stable network may become unstable by adding some delays. Thus, analyses of RNNs with delays is important.

RNNs have been rapidly developed in recent years. Many successful applications of RNNs have been realised. RNNs have found applications in prediction, optimization, image processing, pattern recognisation, association memory, classification, clustering, data mining [143], etc. Hardware circuits implementation of some RNNs are now available, see for examples, [76, 16]. In [76], a hardware of a RNN is implemented to perform digital selection and analogue amplification coexist in a cortex-inspired silicon circuit. Many exciting results on RNNs are still being reported steadily.

## 1.2 Convergence of RNNs

There are many applications of RNNs that relate to the convergence properties of RNNs. Convergence analysis is one of the most important issues of RNNs. Understanding the convergence properties of RNNs is an initial and essential step towards their successful applications. Convergence analysis of

RNNs desires to establish simple, verifiable and weak conditions to guarantee convergence of RNNs. These conditions can be then used to aid the designs of RNNs. This book will mainly focus on the the study of this problem.

Convergence of RNNs can be roughly divided into two classes of convergent behaviours: monostability and multistability. In a monostable RNN, the network has one equilibrium point only. Concepts including global exponential stability, global asymptotic stability, absolute stability, input-to-state stability, etc., describe this behaviour. Generally speaking, monostability requires all the trajectories of a RNN to converge to an equilibrium. Certain existing Lyapunov stability theories could be applied to monostability analysis. However, due to the special nonlinearity presented in many RNNs, the Lyapunov stability theory needs to be further developed to derive good convergence results. Directly using existing theory is also not interesting from a mathematical point of view as this can lead to fairly limited results. In some applications, especially in some optimisation problems, the monostability of RNNs can play a crucial role. Figure 1.2 shows an intuitive illustration of monostable dynamics.

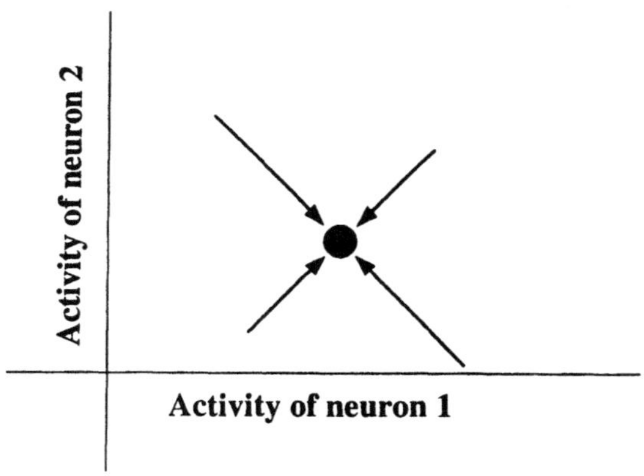


*Figure 1.2.* Monostable dynamics.

Multistability is another convergence behaviour of RNNs. Though monostability of RNNs has successful applications in certain optimization problems, monostable networks are computationally restrictive. They cannot deal with important neural computations such as those necessary in decision making, where multistable networks become necessary. Multistable networks are different from monostable networks, and their analysis also merit different treatment. In a multistable network, stable and unstable equilibrium may coexist in the network. Figure 1.3 shows an intuitive illustration of multistable dynamics. In this figure, the two black dots are stable equilibrium points. The white dot is an unstable equilibrium point.

It must be pointed out that the existence of unstable equilibrium points in a multistable network does not affect its applications. In practice, only stable

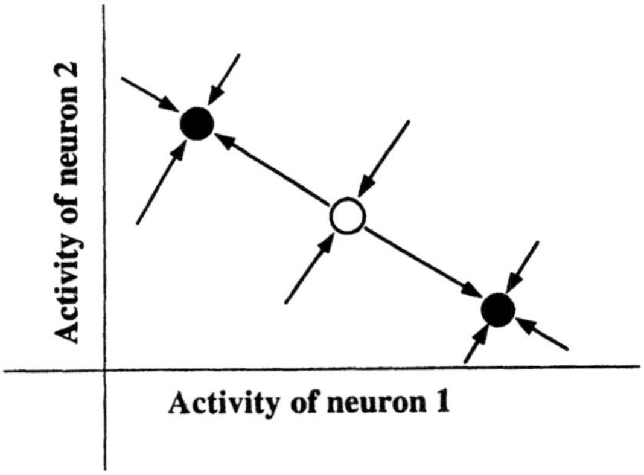


*Figure 1.3.* Multistable dynamics.

states can be observed in a network, because even infinitesimal amounts of noise can lead to divergence from an unstable steady state. In certain applications, such as the Winner-take-all problems, the existence of unstable equilibrium points are essential for digital constraints on selection.

Monostability and multistability analyse constitute the main topics of this book. Besides, the existence and dynamic properties of periodic oscillations of RNNs are also studied.

This book mainly adopt three methods: Lyapunov method, inequalities analysis method, and energy functions method to study the convergence of RNNs. The book desires to seek convergence conditions for RNNs, which are simple, verifiable, weak, and not parallel to existing conditions in mathematical theories. This will be achieved by deeply analyzing the nonlinear characteristic of RNNs.

## 1.3 Outline of the Book

The rest of this chapter is organized as follows. In Section 2, some frequently used notations and terminologies are given. In Section 3, the general method of energy functions is discussed. Energy functions play important roles in multistability analysis.

In Chapter 2, the convergence of Hopfield RNNs described by

$$C_i \frac{du_i(t)}{dt} = -\frac{u_i(t)}{R_i} + \sum_{j=1}^{n} T_{ij} g_j(u_j(t)) + I_i, (i = 1, \cdots, n)$$

is discussed. Complete convergence, local stability as well as global stability are considered.

In Chapter 3, the convergence of three types of Cellular Neural Networks (CNNs) is studied. The three types of CNNs are: standard CNNs described by

$$\dot{x}(t) = -x(t) + f(x(t)) + h,$$

CNNs with constant delays described by

$$\dot{x}(t) = -x(t) + Af(x(t)) + Bf(x(t - \tau(t))) + h,$$

and CNNs with infinite delay described by

$$\begin{aligned} \frac{dx_i(t)}{dt} &= \sum_{j=1}^{n} [a_{ij} f(x_j(t)) + b_{ij} f(x_j(t - \tau_{ij}(t)))] \\ &\quad + \sum_{j=1}^{n} c_{ij} \int_{-\infty}^{t} k_{ij}(t-s) f(x_j(s)) ds \\ &\quad - x_i(t) + I_i, \quad t \geq 0, (i = 1, \cdots, n). \end{aligned}$$

The convergence analyse includes exponential convergence, estimate of exponential convergence rates, complete stability, and global asymptotic stability.

In Chapter 4, the multistability of RNNs with unsaturating piecewise linear transfer functions is studied. The model of this network is described by

$$\dot{x}_i(t) = -x_i(t) + \sum_{j=1}^{n} w_{ij} \sigma(x_j(t)) + h_i, \quad (i = 1, \cdots, n),$$

where the activation function $\sigma$ is defined as

$$\sigma(s) = \max\{0, s\}, \quad s \in R,$$

which is an unsaturating piecewise linear function. This chapter deals fully with three basic properties of a multistable network: boundedness, global attractivity and complete convergence. The stability conditions are based on using local inhibition on connections among neurons.

In Chapter 5, the Lotka-Volterra RNNs (LV RNNs) with delays are considered. The model of LV RNNs with delays is described by

$$\dot{x}_i(t) = x_i(t) \left[ h_i - x_i(t) + \sum_{j=1}^{n} \left( a_{ij} x_j(t) + b_{ij} x_j(t - \tau_{ij}(t)) \right) \right],$$

where $i = 1, \cdots, n$. Both multistability and monostability are analysed. Conditions which allow networks to possess multistability are obtained. Attractors are estimated. By assuming the delays are in $L^2$ space and the symmetry on synaptic connection weights, complete convergence is proven.

In Chapter 6, general delayed RNNs are studied under the assumption that the activation functions are global Lipshictz functions (GLF) are studied. A function $g$ is called GLF, if it is a monotonically increasing and nonconstant function, and it satisfies

$$0 \leq \frac{g(\xi) - g(\eta)}{\xi - \eta} \leq l,$$

for all $\xi, \eta \in R$ and $\xi \neq \eta$, where $l > 0$ is a constant. Global stability of RNNs with constant delays or variable delays are studied. Absolute periodicity and absolute stability of RNNs with delays are also presented in this chapter.

In Chapter 7, the convergence of special types of RNNs is discussed. A detailed convergence analysis is given for a model of RNNs described by

$$\dot{X}(t) = X(t)^T X(t) A X(t) - X(t)^T A X(t) X(t).$$

This model of RNNs can be used for extracting some eigenvectors and eigenvalues of any symmetric matrix. In this chapter, a concept of Input-to-State Convergence (ISC) is proposed to be applied to a model of RNNs with variable inputs described by

$$\frac{dx_i(t)}{dt} = \sum_{j=1}^{n} a_{ij} g_j[x_j(t)] + I_i(t), (i = 1, \cdots, n).$$

Convergent conditions for ISC are established.

In Chapter 8, discrete time RNNs are studied. Multistability is discussed for discrete RNNs with unsaturating piecewise linear transfer functions which are described by

$$x_i(k+1) = \sum_{j=1}^{n} w_{ij} \sigma(x_j(k)) + h_i, \quad (i = 1, \cdots, n).$$

A model of discrete time RNNs described by

$$\begin{cases} x_i(k+1) = \dfrac{(c+\beta)(x_i(k)+\gamma)^\alpha}{\sum\limits_{j=1}^{n}(x_j(k)+\gamma)^\alpha} - \gamma \\ y_i(k) = h(x_i(k)), (i = 1, \cdots, n) \end{cases}$$

is proposed to deal with the winner-take-all problem. The dynamics of this RNN model is presented in details and extensive simulations are given to show the networks performance.

## 2. Some Notations and Terminologies

In this section, some frequently used notations and terminologies of this book are given. Some basic properties are also discussed.

Denote throughout this book,

$$\delta_{ij} = \begin{cases} 1, & i = j \\ 0, & i \neq j \end{cases}$$

for $i, j = 1, \cdots, n$.

The symbol $D^+$ denotes the *upper righthand Dini derivative*. For any continuous function $f : R \to R$, the *upper righthand Dini derivative* of $f(t)$ is defined by

$$D^+ f(t) = \lim_{\theta \to 0^+} \sup \frac{f(t+\theta) - f(t)}{\theta}.$$

It is easy to see that if $f(t)$ is locally Lipschitz then $|D^+ f(t)| < +\infty$.

Let $A$ be a real $n \times n$ matrix, denote through this book $A > 0$, $A < 0$, $A \geq 0$, $A \leq 0$ by $A$ is positive definite, negative definite, semi-positive definite, semi-negative definite, respectively.

A matrix $\Omega = (\omega_{ij})_{n \times n}$ is an *M-matrix*, if $\omega_{ii} \geq 0 (i = 1, \cdots, n)$, $\omega_{ij} \leq 0 (i \neq j; i, j = 1, \cdots, n)$, and the real part of each eigenvalue of $\Omega$ is nonnegative. $\Omega$ is a nonsingular *M-matrix*, if $\Omega$ is both an *M-matrix* and nonsingular. The matrix $\Omega$ is a nonsingular *M-matrix*, if and only if, there exist constants $\alpha_i > 0 (i = 1, \cdots, n)$ such that [20]

$$\alpha_i \omega_{ii} + \sum_{j=1, j \neq i}^{n} \alpha_j \omega_{ij} > 0, (i = 1, \cdots, n).$$

or

$$\alpha_j \omega_{jj} + \sum_{i=1, i \neq j}^{n} \alpha_i \omega_{ij} > 0, (j = 1, \cdots, n).$$

Let $g(t)$ be a continuous function which is defined on $[0, +\infty)$, the upper limit of $g(t)$ is defined as

$$\lim_{t \to +\infty} \sup g(t) = \lim_{t \to +\infty} \left[ \sup_{\theta \geq t} g(\theta) \right].$$

Clearly, if $g(t)$ is upper bounded, its upper limit exists. If

$$\lim_{t \to +\infty} \sup g(t) = \xi < +\infty,$$

then for any $\epsilon > 0$, there exists a $T \geq 0$ such that

$$g(t) \leq \xi + \epsilon$$

for $t \geq T$. Given any $\tilde{t} \geq 0$, there always exists a $\bar{t} > \tilde{t}$ such that

$$\xi - \epsilon \leq g(\bar{t}).$$

## 3. Energy Functions Method for Convergence Analysis

The energy functions method is an important method for multistability analysis of RNNs. Besides, energy functions are very useful in applications of RNNs such as optimizations, association memory, etc. This section provides some basic discussions on the application of energy functions method to convergence analysis of continuous time RNNs.

### 3.1 Energy Functions

Consider a neural network described by the following general differential equation

$$\dot{x}(t) = f(x(t)) \tag{1.1}$$

for $t \geq 0$, where $x \in R^n$, $f$ is assumed to be sufficiently smooth so that to guarantee the existence and uniqueness of the trajectories of (1.1). Let $x_0 \in R^n$, we use $x(t, x_0)$ to denote the trajectory of (1.1) starting from $x_0$.

DEFINITION 1.1 *A point $x^* \in R^n$ is called an equilibrium point of (1.1) if $f(x^*) = 0$. Denote by $\mathcal{E}$ the set of all equilibrium points of (1.1).*

DEFINITION 1.2 *The network (1.1) is called bounded, if every trajectory is bounded.*

DEFINITION 1.3 *Let $x(t, x_0)$ be a trajectory of (1.1). A point $x^\dagger$ is called an $\omega$-limit point of $x(t, x_0)$, if there exists a time sequence $t_k$ with $t_k \to +\infty$ as $k \to +\infty$, such that*

$$\lim_{k \to +\infty} x(t_k, x_0) = x^\dagger.$$

*Denote by $\Omega(x_0)$ the set of all $\omega$-limit points of $x(t, x_0)$ and call $\Omega(x_0)$ the $\omega$-limit set of $x(t, x_0)$.*

LEMMA 1.1 *[77] If $x(t, x_0)$ is bounded, then $\Omega(x_0)$ must be a connected set.*

DEFINITION 1.4 *The network (1.1) is called completely stable, if each trajectory of (1.1) converges to an equilibrium point.*

DEFINITION 1.5 *An equilibrium point $x^*$ of (1.1) is said to be Lyapunov stable, if for any $\epsilon > 0$, there exists a $\delta > 0$ such that $\|x_0 - x^*\| \leq \delta$ implies that*

$$\|x(t, x_0) - x^*\| \leq \epsilon$$

*for all $t \geq 0$.*

DEFINITION 1.6 *An equilibrium point $x^*$ of (1.1) is said to be Lyapunov asymptotically stable, if $x^*$ is Lyapunov stable, and there exist a $\eta > 0$ such that*

$$\lim_{t \to +\infty} x(t, x_0) = x^*$$

*for all $x_0$ satisfies that $\|x_0 - x^*\| \leq \eta$.*

Lyapunov stability and complete stability are two different concepts. A neural network is completely stable does not imply that every equilibrium point of the network is Lyapunov stable. In a complete stable network, stable and unstable equilibrium points may coexist. To make this more clear, consider an example to illustrate this point. Consider the following one dimensional Hopfield neural network

$$\dot{x}(t) = -\frac{3}{5\ln 2}x(t) + g(x(t)), \quad t \geq 0 \tag{1.2}$$

where

$$g(s) = \frac{e^s - e^{-s}}{e^s + e^{-s}}, \quad s \in R.$$

This network has three equilibrium points: 0, $\ln 2$ and $-\ln 2$. Given any $x_0 \in R$, let $x(t, x_0)$ be the trajectory of (1.2) starting from $x_0$. It is easy to see that

$$x(t, x_0) \begin{cases} \rightarrow \ln 2, & \text{if } x_0 > 0 \\ \rightarrow -\ln 2, & \text{if } x_0 < 0 \end{cases}$$

as $t \rightarrow +\infty$, and $x(t, x_0) \equiv 0$ if $x_0 = 0$. Clearly, this network is completely stable since every trajectory converges to an equilibrium. However, the equilibrium point 0 is not Lyapunov stable whereas $\ln 2$ and $-ln2$ are Lyapunov asymptotic stable equilibrium points. Figure 1.4 shows the trajectories of (1.2).

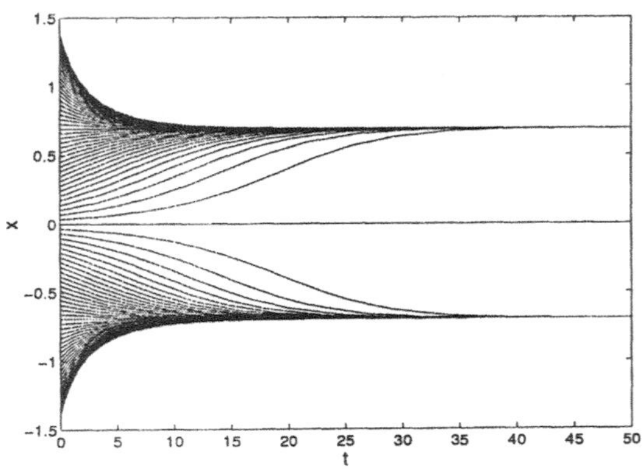


*Figure 1.4.* Lyapunov stable and unstable equilibrium points may coexist in a complete stable network.

DEFINITION 1.7 *A continuous function $E : R^n \to R$ is called an energy function of (1.1), if the derivative of $E$ along the trajectories of (1.1) satisfies*

$$\begin{cases} \dot{E}(x(t)) \leq 0 \\ \dot{E}(x(t)) = 0 \quad \textit{if and only if} \quad \dot{x}(t) = 0 \end{cases}$$

*for all $t \geq 0$.*

It should be noted that an energy function may take positive or negative values. This is different from Lyapunov functions. Usually, a Lyapunov function is required to be positive or nonnegative. Generally speaking, Lyapunov functions are suitable for analyze stability of some equilibrium point, while energy functions are suitable for multistability analyze such those as complete stable network. The follows is a fundamental theorem for energy functions.

THEOREM 1.1 *Suppose the network (1.1) is bounded and has an energy function $E(x)$. Then, for all $x_0 \in R^n$, it holds that*

$$\Omega(x_0) \subset \mathcal{E}.$$

*If, in particular, the equilibrium points are isolated, then the network (1.1) is completely stable.*

*Proof*: Given any $x_0 \in R^n$. By condition of Theorem 1.1, $x(t, x_0)$ is bounded. Then, there exists a bounded closed set $S \subset R^n$ such that $x(t, x_0) \subset S$ for all $t \geq 0$. Let $\Omega(x_0)$ be the $\omega$-limit set of $x(t, x_0)$. Clearly, $\Omega(x_0) \neq \phi$, and $\Omega(x_0) \subset S$. By assumption of the theorem, the network has an energy function $E(x)$. Since, $\dot{E}(x(t, x_0) \leq 0$ for $t \geq 0$, and $x(t, x_0) \subset S$, then $E(x(t, x_0))$ is a monotone decrease bounded function, thus, there must exist a constant $E_0$ such that

$$\lim_{t \to +\infty} E(x(t, x_0)) = E_0.$$

For any $y \in \Omega(x_0)$, let $x(t, y)$ be the trajectory of (1.1) passing through $y$. Clearly, $x(t, y) \in \Omega(x_0)$ for every $t \geq 0$. Since $E$ is continuous, then $E(x(t, y)) = E_0$ for all $t \geq 0$. Moreover,

$$\dot{E}(x(t, y)) = 0, \quad t \geq 0.$$

Then, it follows that $\dot{x}(t, y) = 0$ for all $t \geq 0$. This implies that $x(t, y) \equiv y$ is an equilibrium point, i.e, $y \in \mathcal{E}$. Since $y$ is arbitrarily chosen in $\Omega(x_0)$, then all the points of $\Omega(x_0)$ are equilibrium points, i.e.,

$$\Omega(x_0) \subset \mathcal{E}.$$

If, in particular, the equilibrium points are isolated, then the points of $\Omega(x_0)$ are also isolated. By Lemma 1.1, every $\omega$-limit set is connected, then $\Omega(x_0)$ must contain one equilibrium point $x^*$ only. Thus, $x(t, x_0)$ converges to $x^*$, i.e., the network (1.1) is completely stable. This completes the proof.

Theorem 1.1 shows that if the equilibria of a network are isolated, then it is completely stable. If the equilibria of a network are not isolated, by Theorem 1.1, it can only confirm that $\Omega(x_0) \subset \mathcal{E}$ for any $x_0 \in R^n$. A problem is that whether the network is still completely stable or not? Currently, no confirmed answer to this problem is available. Further research to this problem is required. To some particular networks, in the case that equilibria are not isolated, complete stability can be proven by plus some additional analysis besides energy functions. This will be shown in the next fewer chapters of this book.

The boundedness condition in above theorem is important. If this condition is removed, a network may not be completely stable. For example, consider the simple one dimensional equation

$$\dot{x}(t) = x(t), \quad t \geq 0. \tag{1.3}$$

Define

$$E(x) = \frac{1}{1 + x^2}.$$

Clearly, $E(x)$ is bounded and the derivative of $E$ along the trajectories of (1.3) satisfies

$$\dot{E}(x(t)) = -\frac{2x(t)}{(1 + x^2(t))^2}\dot{x}(t) = -\frac{2}{(1 + x^2(t))^2}(\dot{x}(t))^2 \leq 0.$$

Clearly, $E(x)$ is an energy function. However, if $x_0 \neq 0$, the trajectory $x(t, x_0) = x_0 e^t \to +\infty$. This shows that (1.3) is not completely stable.

## 3.2 Relationships Between Minimums of Energy Functions and Equilibria of Networks

From application point of view, if a network posses an energy function, it is expected that each equilibrium of the network is just a minimum of the energy function. However, this is not always true. Consider again the network (1.2).

This network has three equilibrium points 0, $\ln 2$ and $-\ln 2$. Clearly, the equilibria are isolated. This network possess a Hopfield energy function which is given by

$$E(x) = -\frac{1}{2}g^2(x) + \frac{3}{5\ln 2}\int_0^{g(x)} g^{-1}(\theta)d\theta.$$

Thus, the network is completely stable.

It is easy to see that

$$\frac{dE(x)}{dx} = -4\frac{e^x - e^{-x}}{(e^x + e^{-x})^3} + \frac{3}{5\ln 2}x$$

and

$$\frac{d^2E(x)}{dx^2} = \frac{3}{5\ln 2} - 4\frac{(e^x + e^{-x})^2 - 3(e^x - e^{-x})^2}{(e^x + e^{-x})^4}.$$

Then,

$$\frac{dE(0)}{dx} = 0, \quad \frac{d^2E(0)}{dx^2} < 0.$$

This shows that the equilibrium point 0 is a maximum of the energy function.

By similar calculations, it can be seen that $\ln 2$ and $-\ln 2$ are minimas of the energy function.

A natural question now arises. Under what conditions can an equilibrium point be a minimum of an energy function? What are the relationships between the equilibria and the minimas of energy functions? Next, such problems are addressed.

THEOREM 1.2 *Suppose that the network (1.1) is bounded and has a energy function $E(x)$. Let $x^*$ be an equilibrium point of (1.1). Then, $x^*$ is a minimum of $E(x)$ if and only if there exists a sufficiently small neighborhood $N(x^*)$ such that for any $x_0 \in N(x^*)$, it holds that $x(t, x_0) \to x_0^*$ as $t \to +\infty$, and $E(x_0^*) \geq E(x^*)$.*

*Proof*: Necessity. Suppose that $x^*$ is an equilibrium point of (1.1) and it is also a minimum of the energy function $E(x)$. If the assertion of the theorem is not true, then, there must exist a sequence $x_k \to x^*(k = 1, 2, 3, \cdots)$ such that

$$\begin{cases} E(x_k) \geq E(x^*) \\ \lim_{t \to +\infty} x(t, x_k) = x_k^* \\ E(x_k^*) < E(x^*). \end{cases}$$

Since the energy function is decreasing along each trajectory, there must exist $t_k \to 0(k \to +\infty)$ such that

$$\begin{cases} \lim_{t \to +\infty} x(t_k, x_k) = x^* \\ E(x(t_k, x_k) < E(x^*). \end{cases}$$

Clearly, $x^*$ cannot be a minimum of $E(x)$. This contradicts the assumption.

Sufficiency. Given any $x_0 \in N(x^*)$, by Theorem 1.1, the network (1.1) is completely stable. Then, $x(t, x_0)$ converges to an equilibrium point $x^\perp$. Since $E(x(t, x_0)$ is decreasing, it follows that

$$E(x_0) \geq E(x(t, x_0) \geq E(x^\perp) \geq E(x^*).$$

As $x_0$ is arbitrary in $N(x^*)$, then $x^*$ must be a minimum of $E(x)$. The proof is completed.

COROLLARY 1.1 *Suppose the network (1.1) is bounded and has an energy function $E(x)$. Then, any local stable equilibrium point is a local minimum of $E(x)$.*

When networks are applied to optimisation problems, the global minimum of energy functions is an important concern.

THEOREM 1.3 *Suppose the network (1.1) is bounded and has an energy function $E(x)$. Let $x^*$ be an equilibrium point of (1.1). Then, $x^*$ is a global minimum of $E(x)$ if and only if for any other equilibrium $x^{\perp}$, it holds that $E(x^{\perp}) \geq E(x^*)$.*

*Proof*: Necessity. Let $x^*$ be an equilibrium of the network and global minimum of $E(x)$. If the conclusion of the theorem is not true, then there must have another equilibrium $x^{\perp}$ such that $E(x^{\perp}) < E(x^*)$. This contradict with that $x^*$ is a global minimum of $E(x)$.

Sufficiency. Given any $x_0 \in R^n$, by Theorem 1.2, the network (1.1) is completely stable, then $x(t, x_0)$ converges to an equilibrium point $x^{\perp}$. Since $E(x(t, x_0))$ is decreasing, it follows that

$$E(x_0) \geq E(x(t, x_0)) \geq E(x^{\perp}) \geq E(x^*).$$

for all $t \geq 0$. This shows that $x^*$ is a global minimum of the energy function $E(x)$. The proof is completed.

COROLLARY 1.2 *Suppose the network (1.1) is bounded and has an energy function $E(x)$. If the network (1.1) has only one equilibrium point, then it must be the global minimum of $E(x)$.*

It is worth able to point out that if an energy function of a network has a global minimum, it does not imply that the network has only one equilibrium point. For example, consider the following linear network

$$\dot{x}(t) = -\sigma x(t) + Ax(t), \quad t \geq 0, \tag{1.4}$$

where $x \in R^n$, $A$ is a $n \times n$ matrix, $\sigma > 0$ is the largest eigenvalue of $A$.

Since all the eigenvalues of $A - \sigma I$ are non-positive, then the network is bounded. Constructing a continuous function

$$E(x) = -\frac{1}{2}x^T Ax + \frac{\sigma}{2}x^T x.$$

Then,

$$\dot{E}(x(t)) = -\|\dot{x}(t)\|^2, \quad t \geq 0.$$

Clearly, $E(x)$ is an energy function. It is easy to see that the set of equilibrium points is the eigen-subspace $V_\sigma$ of $A$ corresponding to the largest eigenvalue

$\sigma$. For any $x^* \in V_\sigma$, it holds that $E(x^*) = 0$. Using Theorem 1.3, each point of $V_\sigma$ is a global minimum of $E(x)$. This shows that the energy function $E(x)$ has a lot of global minimum since $V_\sigma$ is a non-countable set.

As mentioned above, the equilibria of (1.4) are not isolated. Is this network completely stable? Theorem 1.1 cannot completely address this problem by an energy function. Next, another way will be used to solve this problem.

Since $A$ is symmetric, then there exists an orthonormal basis of $R^n$ composed by eigenvectors of $A$. Let $\lambda_i (i = 1, \cdots, n)$ be eigenvalues of $A$ and $S_i (i = 1, \cdots, n)$ be the corresponding eigenvectors that compose an orthonormal basis of $A$. Then, for any $x_0 \in R^n$, the trajectory $x(t) = x(t, x_0)$ can be represented as $x(t) = \sum_{i=1}^{n} z_i(t) S_i$, where $z_i(t)(i = 1, \cdots, n)$ are some differentiable functions. Then, from (1.4), it follows that

$$\dot{z}_i(t) = -(\sigma - \lambda_1) z_i(t), (i = 1, \cdots, n)$$

for $t \geq 0$. Thus,

$$z_i(t) = z_i(0) e^{-(\sigma - \lambda_1)t}, (i = 1, \cdots, n)$$

for $t \geq 0$. Without lose of generality, assume that $\lambda_1$ is the largest eigenvalue of $A$, i.e., $\lambda_1 = \sigma$. Suppose the multiplicity of $\sigma$ is $k$, then, clearly

$$z_i(t) = \begin{cases} z_i(0), (i = 1, \cdots, k) \\ \\ \rightarrow 0, \text{as } t \rightarrow +\infty. \end{cases}$$

Thus,

$$x(t) \rightarrow \sum_{i=1}^{k} z_i(0) S_i \in V_\sigma$$

as $t \rightarrow +\infty$. This shows that every trajectory of (1.4) converges to an equilibrium point. Then, the linear network (1.4) is completely stable.

## 4. Conclusions

This chapter gives simple introductions to RNNs and convergence concepts of RNNs as well as outline of the book. The method of energy functions for convergence is discussed. Fundamental theorems of energy functions related to convergence of RNNs are proved. Energy functions are frequently used in RNNs to some applications such as optimizations and association memory. Compared to Lyapunov functions, energy functions are more suitable for the multistability analysis of RNNs. However, the method of energy functions cannot solve all related problems. Further research to complete the method of energy functions is required. Other recent results on energy functions and complete stability can be found in [58, 59, 124].

Chapter 2

# HOPFIELD RECURRENT NEURAL NETWORKS

## 1. Introduction

In 1982, Hopfield proposed a model of neural networks [84], which used two-state threshold "neurons" that followed a stochastic algorithm. This model explored the ability of a network of highly interconnected "neurons" to have useful collective computational properties, such as content addressable memory. However, the model is based on McCulloch-Pitts neurons that are different from real biological neurons and also from the realistic functioning of simple electric circuits. Real neurons have continuous input-output relations and integrative time delays due to capacitance. To overcome such problems, in 1984, Hopfield proposed another continuous time recurrent neural network model with a graded response. It is described by a set of differential equations. This deterministic system has collective properties very close to the earlier stochastic model. Today, this model is well known as the Hopfield model of RNNs and it has found wide applications in various optimisation problems [65, 22, 107, 182], associative memories, engineering problems, satellite broadcast scheduling problems [64, 4], graph partition [190], stereo vision [145], multiuser detector [101], fault detection and isolation [172], affine invariant matching [109], pattern sequence recognition [105], classification [26], etc. The contributions of Hopfield RNN model to the field of neural networks cannot be over-emphasised. In fact, it is the outstanding work of Hopfield that has rekindled research interests in the neural networks from both scientists and engineers.

The model of Hopfield RNNs is described by

$$\begin{cases} C_i \dfrac{du_i(t)}{dt} = -\dfrac{u_i(t)}{R_i} + \displaystyle\sum_{j=1}^{n} T_{ij} v_i(t) + I_i \\ \\ v_i(t) = g_i(u_i(t)), (i = 1, \cdots, n) \end{cases} \tag{2.1}$$

for $t \geq 0$. In this model, $u_i$ is the state of neuron $i$, $v_i$ is output of neuron $i$ with input output relation $v_i = g_i(u_i)$. Each $g_i$ is a sigmoid function. $v_i$ can be viewed as the short-term average of the firing rate of the neuron $i$. $T_{ij}v_j$ represents the electrical current input to neuron $i$ due to the present potential of neuron $i$, and $T_{ij}$ thus represents the synaptic connection strength between neurons $i$ and $j$. $T_{ij}$ can be both positive or negative, representing excitation or inhibition connection between neurons. $I_i$ is the external constant input to neuron $i$. $C_i > 0$ and $R_i > 0$ are the neuron amplifier input capacitance and resistance, respectively. An electrical circuit that corresponds to the network model (2.1) is shown in Figure 2.1.

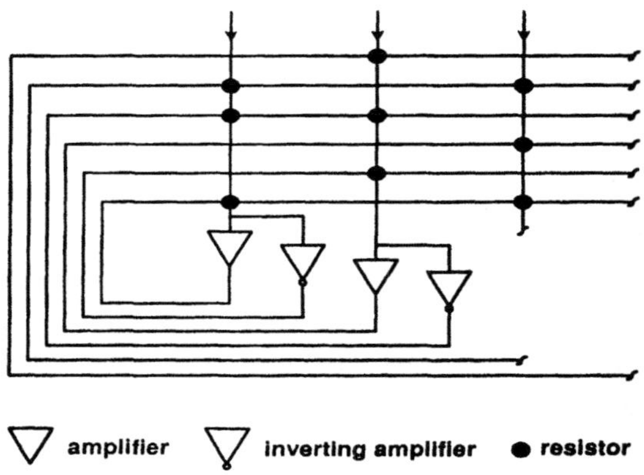


*Figure 2.1.* Electric circuit of Hopfield RNNs.

In the Hopfield RNNs, the activation functions $g_i (i = 1, \cdots, n)$ are continuous and differentiable nonlinear sigmoid functions satisfying the following conditions:

$$\begin{cases} \lim_{s \to \pm\infty} g_i(s) = \pm 1 \\ |g_i(s)| \leq 1, \quad s \in R \\ 0 < \dot{g}_i(s) < \dot{g}_i(0), \quad s \in R - \{0\} \\ \lim_{s \to \pm\infty} \dot{g}_i(s) = 0. \end{cases} \tag{2.2}$$

Typical examples of sigmoid functions are

$$tanh(s) = \frac{e^s - e^{-s}}{e^s + e^{-s}}, \quad \frac{1 - e^{-s}}{1 + e^{-s}}, \quad \frac{2}{\pi} tan^{-1}\left(\frac{\pi}{2}s\right), \quad \frac{1}{1 + e^{-s}}.$$

Figure 2.2 shows the sigmoid function $g(s) = 1/(1 + e^{-s})$ and the derivative $\dot{g}(0)$.

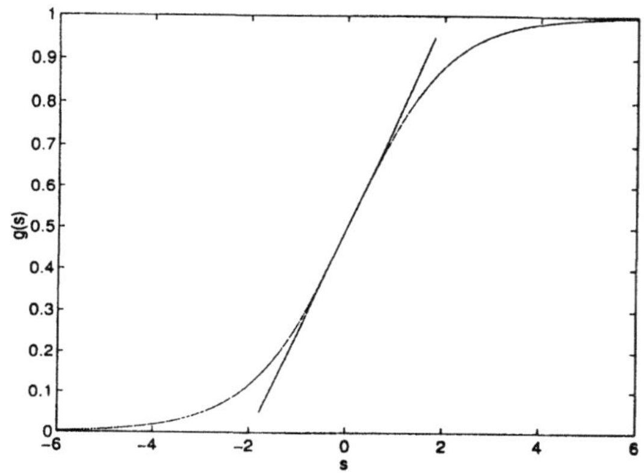


*Figure 2.2.* Sigmoid function $g(s) = 1/(1 + e^{-s})$ and $\dot{g}(0)$.

This chapter is organised as follows. In Section 2, the equilibria analysis is given. Complete stability analysis using Hopfield energy functions is discussed in Section 3. Exponential convergence is studied in Section 4. In Section 5, the conclusion to the chapter is presented.

## 2. Equilibria Analysis

A point $u^* = (u_1^*, \cdots, u_n^*)^T \in R^n$ is called an *equilibrium* of (2.1) if it satisfies

$$-\frac{1}{R_i}u_i^* + \sum_{j=1}^{n} T_{ij}g_j(u_j^*) + I_i \equiv 0, (i = 1, \cdots, n) \qquad (2.3)$$

for $i = 1, \cdots, n$.

Equilibria analysis aims to investigate the conditions for the equilibria to exist, the uniqueness of the equilibria, and the locations of equilibria. To completely answer these problems is not easy in general since essentially this problem belongs to a class of algebraic curve problems which have not been solved yet.

THEOREM 2.1 *The network of (2.1) has at least one equilibrium point.*

*Proof:* Define a mapping $\varphi : R^n \to R^n$ as

$$\varphi(u_i) = R_i \left[ \sum_{j=1}^{n} T_{ij}g_j(u_j) + I_i \right]$$

for $i = 1, \cdots, n$. Since, from (2.2), $|g_i(\cdot)| \leq 1$, then

$$|\varphi(u_i)| \leq R_i \left[ \sum_{j=1}^{n} |T_{ij}| + |I_i| \right]$$

for $i = 1, \cdots, n$. Obviously, $\varphi$ maps any bounded closed set into a bounded closed set. Then by the Brouwer fixed point theorem [83], the map $\varphi$ has at least one fixed point $u^* = (u_1^*, \cdots, u_i^*)^T \in R^n$, with $\varphi(u^*) = u^*$, *i.e.*,

$$-\frac{1}{R_i}u_i^* + \sum_{j=1}^{n} T_{ij}g_j(u_j^*) + I_i = 0, (i = 1, \cdots, n).$$

giving an equilibrium. The proof is completed.

THEOREM 2.2 *If*

$$\left[-\frac{1}{\dot{g}_i(0)R_i}\delta_{ij} + \frac{T_{ij} + T_{ji}}{2}\right]_{n\times n} \leq 0,$$

*the network (2.1) has a unique equilibrium.*

*Proof*: By Theorem 2.1, the network (2.1) has at least one equilibrium. Thus, to complete the proof, it is sufficient to prove the uniqueness. Suppose there are two equilibrium points $u^*$ and $u^{\perp}$. Then, from (2.3), it follows that

$$\frac{1}{R_i}\left(u_i^* - u_i^{\perp}\right) = \sum_{j=1}^{n} T_{ij}\left[g_j(u_j^*) - g_j(u_j^{\perp})\right].$$

Then,

$$\begin{aligned} & \frac{1}{R_i}\left[u_i^* - u_i^{\perp} - \frac{g_j(u_j^*) - g_j(u_j^{\perp})}{\dot{g}_i(0)}\right] \\ = & \sum_{j=1}^{n}\left(-\frac{\delta_{ij}}{\dot{g}_i(0)R_i} + T_{ij}\right)\left[g_j(u_j^*) - g_j(u_j^{\perp})\right]. \end{aligned} \tag{2.4}$$

Denote

$$\Upsilon = \sum_{i=1}^{n}\frac{1}{R_i}\left[g_i(u_i^*) - g_i(u_i^{\perp})\right] \cdot \left[u_i^* - u_i^{\perp} - \frac{g_i(u_i^*) - g_i(u_i^{\perp})}{\dot{g}_i(0)}\right].$$

Clearly, $\Upsilon \geq 0$. From (2.4), it follows that

$$\begin{aligned} \Upsilon &= \sum_{i=}^{n}\sum_{j=1}^{n}\left(-\frac{\delta_{ij}}{\dot{g}_i(0)R_i} + T_{ij}\right)\left[g_i(u_i^*) - g_i(u_i^{\perp})\right] \cdot \left[g_j(u_j^*) - g_j(u_j^{\perp})\right] \\ &= \sum_{i=}^{n}\sum_{j=1}^{n}\left(-\frac{\delta_{ij}}{\dot{g}_i(0)R_i} + \frac{T_{ij} + T_{ji}}{2}\right)\left[g_i(u_i^*) - g_i(u_i^{\perp})\right] \\ &\quad \times \left[g_j(u_j^*) - g_j(u_j^{\perp})\right] \\ &\leq 0. \end{aligned} \tag{2.5}$$

It will be shown that $u_i^* = u_i^{\perp}(i = 1, \cdots, n)$. If not, there must exist some $i$ such that $u_i^* \neq u_i^{\perp}$. Then,

$$\begin{aligned} \Upsilon &\geq \frac{1}{R_i}\left[g_i(u_i^*) - g_i(u_i^{\perp})\right] \cdot \left[u_i^* - u_i^{\perp} - \frac{g_i(u_i^*) - g_i(u_i^{\perp})}{\dot{g}_i(0)}\right] \\ &= \frac{1}{R_i}\left[g_i(u_i^*) - g_i(u_i^{\perp})\right]^2 \cdot \left[\frac{u_i^* - u_i^{\perp}}{g_i(u_i^*) - g_i(u_i^{\perp})} - \frac{1}{\dot{g}_i(0)}\right] \\ &= \frac{1}{R_i}\left[g_i(u_i^*) - g_i(u_i^{\perp})\right]^2 \cdot \left[\frac{1}{\dot{g}_i(\theta)} - \frac{1}{\dot{g}_i(0)}\right] \end{aligned}$$

where $\theta$ is located between $u_i^*$ and $u_i^{\perp}$. Since $u_i^* \neq u_i^{\perp}$, from (2.2), it must hold that $\theta \neq 0$. Thus, $\Upsilon > 0$. This contradicts (2.5). It follows that $u_i^* = u_i^{\perp}(i = 1, \cdots, n)$. This completes the proof.

The condition of Theorem 2.2 can be relaxed by introducing some parameters as follows.

THEOREM 2.3 *If there exist constants $\gamma_i > 0(i = 1, \cdots, n)$ such that*

$$\left[-\frac{\gamma_i}{\dot{g}_i(0)R_i}\delta_{ij} + \frac{\gamma_i T_{ij} + \gamma_j T_{ji}}{2}\right]_{n\times n} \leq 0,$$

*the network (2.1) has a unique equilibrium.*

*Proof*: Rewrite (2.3) as

$$-\frac{\gamma_i}{R_i}u_i^* + \sum_{j=1}^{n} \gamma_i T_{ij} g_j(u_j^*) + \gamma_i I_i = 0, (i = 1, \cdots, n),$$

for $i = 1, \cdots, n$. Then, Theorem 2.2 is applicable. The proof is completed.

Consider an example

$$\dot{u}(t) = -u(t) + g(u(t)) + I,$$

for $t \geq 0$, where

$$g(s) = \frac{e^s - e^{-s}}{e^s + e^{-s}}.$$

Since

$$\dot{g}(s) = 1 - g(s)^2 < \dot{g}(0) = 1, s \in R - \{0\},$$

clearly, it satisfies the conditions of Theorem 2.2. Thus, this network has only one equilibrium.

Theorem 2.2 and Theorem 2.3 give some sufficient conditions for the uniqueness of equilibrium point. However, in some situations, (2.1) may have more than one equilibrium.

THEOREM 2.4 *The equilibria of the network (2.1) must be isolated.*

*Proof*: If the equilibria are not isolated, then the network (2.1) must have infinite equilibria, i.e., (2.3) have infinite solutions. By (2.3), the matrix

$$\begin{bmatrix} -\frac{u_1}{R_1 g_1(u_1)} + T_{11} & T_{12} & \cdots & T_{1n} \\ T_{21} & -\frac{u_2}{R_2 g_2(u_2)} + T_{22} & \cdots & T_{2n} \\ \vdots & \vdots & \vdots & \\ T_{n1} & T_{n2} & \cdots & -\frac{u_n}{R_n g_n(u_n)} + T_{nn} \end{bmatrix}$$

should be singular in some domain of $R^n$. By the properties of sigmoid functions defined in (2.2), clearly, this is impossible. The proof is completed.

## 3. Complete Stability

In this section, complete stability of Hopfield RNNs will be studied by using Hopfield energy functions. Throughout this section, denote $T = (T_{ij})_{n \times n}$.

THEOREM 2.5 *If the matrix $T$ is symmetric, then the Hopfield network (2.1) is completely stable.*

*Proof*: Firstly, it will be proven that (2.1) is bounded. From (2.1), it follows that

$$u_i(t) = u_i(0) e^{-\frac{t}{C_i R_i}} + \frac{1}{C_i} \int_0^t e^{-\frac{(t-s)}{C_i R_i}} \left( \sum_{j=1}^n T_{ij} g_j(u_j(s)) + I_i \right) ds,$$

for $t \geq 0$ and $(i = 1, \cdots, n)$. Then,

$$\begin{aligned} |u_i(t)| &\leq |u_i(0)| e^{-\frac{t}{C_i R_i}} + \frac{1}{C_i} \int_0^t e^{-\frac{(t-s)}{C_i R_i}} \left| \sum_{j=1}^n T_{ij} g_j(u_j(s)) + I_j \right| ds \\ &\leq |u_i(0)| + R_i \left( \sum_{j=1}^n |T_{ij}| + |I_i| \right), \end{aligned}$$

for $t \geq 0$ and $(i = 1, \cdots, n)$. Thus, (2.1) is bounded.

Constructing the differentiable function

$$E(u) = -\frac{1}{2} \sum_{i=1}^n \sum_{j=1}^n T_{ij} v_i v_j - \sum_{i=1}^n I_i v_i + \sum_{i=1}^n \frac{1}{R_i} \int_0^{v_i} g_i^{-1}(\theta) d\theta, \qquad (2.6)$$

where $v_i = g_i(u_i)(i = 1, \cdots, n)$ and $u = (u_1, \cdots, u_n)^T$.

Calculating the derivative of $E$ along the trajectories of (2.1) using the fact that $T_{ij} = T_{ji}(i, j = 1, \cdots, n)$, it follows that

$$\begin{aligned}
\left.\frac{dE(u(t))}{dt}\right|_{(2.1)} &= -\sum_{j=1}^{n} \frac{dv_i(t))}{dt}\left(-\frac{1}{R_i}u_i(t) + \sum_{j=1}^{n} T_{ij}v_j(t) + I_i\right) \\
&= -\sum_{j=1}^{n} \frac{1}{C_i}\frac{dv_j(t)}{dt} \cdot \frac{du_i(t)}{dt} \\
&= -\sum_{j=1}^{n} d_i \frac{dg_j(u_i(t))}{du_i(t)} \cdot \left(\frac{du_i(t)}{dt}\right)^2 \\
&\leq 0
\end{aligned}$$

for $t \geq 0$. Clearly, $E(u)$ is an energy function. Then, by Theorem 2.4 and Theorem 1.1, network (2.1) is completely stable. The proof is completed.

The energy function defined in (2.6) is called Hopfield energy of (2.1).

COROLLARY 2.1 *Any one of the following conditions guarantee the network (2.1) is complete stable.*
1). $n = 1$;
2). $n = 2$, *and* $a_{12} = a_{21} = 0$ *or* $a_{12} = a_{21} \neq 0$;
3). $n = 3$, $a_{31}a_{23}a_{12} = a_{13}a_{21}a_{32}$, *and* $a_{ij} = a_{ji} = 0$ *or* $a_{ij}a_{ji} \neq 0, (i \neq j; i, j = 1, 2, 3)$.

The symmetry condition in Theorem 2.5 can be relaxed by introducing some parameters.

THEOREM 2.6 *If there exists a diagonal matrix $D > 0$ such that $DT$ is symmetric, then (2.1) is completely stable.*

*Proof*: Denote $D = diag(d_1, \cdots, d_n)$. Rewrite (2.1) as

$$d_i C_i \frac{du_i(t)}{dt} = -\frac{d_i}{R_i}u_i(t) + \sum_{j=1}^{n} d_i T_{ij} g_j(u_j(t)) + d_i I_i, (i = 1, \cdots, n).$$

Then, Theorem 2.5 is applicable and the proof is completed.

## 4. Global Convergence

This section will discuss the global convergence of the Hopfield neural network (2.1). Both global asymptotic stability and global exponential stability will be considered. The exponential convergence rate will be estimated.

## 4.1 Global Asymptotic Stability

This subsection introduce a weak condition for global asymptotic stability of (2.1) which was first reported in [125]. A slight different proof will be given here.

LEMMA 2.1 *It holds that*

$$\frac{g_i(u_i) - g_i(u_i^*)}{u_i - u_i^*} < \dot{g}_i(0), (i = 1, \cdots, n)$$

*for all* $u_i \neq u_i^*$.

The proof follows directly from the properties of (2.2).

THEOREM 2.7 *If*

$$\left[ -\frac{1}{\dot{g}_i(0) R_i} \delta_{ij} + \frac{T_{ij} + T_{ji}}{2} \right]_{n \times n} \leq 0,$$

*the network (2.1) is globally asymptotically stable.*

*Proof*: Using Theorem 2.2, the network (2.1) has only one equilibrium $u^* = (u_1^*, \cdots, u_n^*)^T$. Denote $u = (u_1, \cdots, u_n)^T$. Define a Lyapunov function

$$V(u) = \sum_{i=1}^{n} C_i \int_{u_i^*}^{u_i} [g_i(s) - g_i(u_i^*)] \, ds. \tag{2.7}$$

Clearly, $V(u) \geq 0$ for all $u \in R^n$, and $V(u) = 0$ if and only if $u \equiv u^*$.

Define another function

$$W(u) = \sum_{i=1}^{n} [g_i(u_i) - g_i(u_i^*)] \left[ u_i - u_i^* - \frac{g_i(u_i) - g_i(u_i^*)}{\dot{g}_i(0)} \right].$$

By Lemma 2.1 and (2.2), clearly, $W(u) \geq 0$ for all $u \in R^n$ and $W(u) = 0$ if and only if $u \equiv u^*$.

Calculating the derivative of $V(u)$ along the trajectories of (2.1), it follows that

$$\begin{aligned} \left.\frac{dV(u)}{dt}\right|_{(2.1)} &= -\sum_{i=1}^{n} \frac{1}{R_i} [g_i(u_i(t)) - g_i(u_i^*)] (u_i(t) - u_i^*) \\ &\quad + \sum_{i=1}^{n} \sum_{j=1}^{n} T_{ij} [g_i(u_i(t)) - g_i(u_i^*)] \left[ g_j(u_j(t)) - g_j(u_j^*) \right] \\ &= -\sum_{i=1}^{n} [g_i(u_i(t)) - g_i(u_i^*)] \end{aligned}$$

$$
\begin{aligned}
& \times \left[u_i(t) - u_i^* - \frac{g_i(u_i(t)) - g_i(u_i^*)}{\dot{g}_i(0)}\right] \\
& + \sum_{i=1}^{n}\sum_{j=1}^{n}\left[-\frac{1}{\dot{g}_i(0)R_i}\delta_{ij} + \frac{T_{ij}+T_{ji}}{2}\right][g_i(u_i(t)) - g_i(u_i^*)] \\
& \times \left[g_j(u_j(t)) - g_j(u_j^*)\right] \\
\leq & -\sum_{i=1}^{n}[g_i(u_i(t)) - g_i(u_i^*)] \\
& \times \left[u_i(t) - u_i^* - \frac{g_i(u_i(t)) - g_i(u_i^*)}{\dot{g}_i(0)}\right] \\
= & -W(u(t))
\end{aligned}
$$

for $t \geq 0$. Thus, by applying the well known Lyapunov stability theory, the network (2.1) is globally asymptotically stable. The proof is completed.

## 4.2 Global Exponential Stability

The following lemmas will be useful.

LEMMA 2.2 *It holds that*

$$[g_i(\eta) - g_i(\xi)](\eta - \xi) \leq \dot{g}_i(0)(\eta - \xi)^2,$$

*and*

$$\int_{\xi}^{\eta}[g_i(s) - g_i(\xi)] \leq [g_i(\eta) - g_i(\xi)](\eta - \xi),$$

*for all $\eta, \xi \in R$.*

The proof of Lemma 2.2 follows directly from the properties of sigmoid functions defined in (2.2). Details are omitted.

LEMMA 2.3 *It holds that*

$$\int_{\xi}^{\eta}[g_i(s) - g_i(\xi)]\,ds \geq \frac{1}{2\dot{g}_i(0)}[g_i(\eta) - g_i(\xi)]^2, (i = 1, \cdots, n),$$

*for all $\eta, \xi \in R$.*

*Proof*: Since $g_i(i = 1, \cdots, n)$ are differentiable and $0 < \dot{g}_i(s) \leq \dot{g}_i(0)$ for all $s \in R$, it follows that

$$
\begin{aligned}
\int_{\xi}^{\eta}[g_i(s) - g_i(\xi)]\,ds &= \int_{\xi}^{\eta}\frac{g_i(s) - g_i(\xi)}{\dot{g}_i(s)}d[g_i(s) - g_i(\xi)] \\
&\geq \frac{1}{\dot{g}_i(0)}\int_{\xi}^{\eta}[g_i(s) - g_i(\xi)]\,d[g_i(s) - g_i(\xi)] \\
&= \frac{1}{2\dot{g}_i(0)}[g_i(\eta) - g_i(\xi)]^2.
\end{aligned}
$$

The proof is completed.

THEOREM 2.8 *If*

$$\Pi = \left[ -\frac{1}{\dot{g}_i(0)R_i}\delta_{ij} + \frac{T_{ij} + T_{ji}}{2} \right]_{n\times n} < 0, \tag{2.8}$$

*then, the network (2.1) is globally exponentially stable. Moreover, the network has a unique equilibrium* $u^* = (u_1^*, \cdots, u_n^*)^T \in R^n$*, and given any* $u(0) \in R^n$*, the trajectory* $u(t) = (u_1(t), \cdots, u_n(t))^T$ *of (2.1) starting from* $u(0)$ *satisfies*

$$\begin{aligned} |u_i(t) - u_i^*| \quad &\leq \quad \|u(0) - u^*\| \left[ e^{-\frac{t}{R_i C_i}} + \frac{2R_i}{2 - \sigma R_i C_i} \sum_{j=1}^{n} |T_{ij}| \right. \\ &\qquad \left. \times \sqrt{\frac{2\dot{g}_i(0)}{C_i} \sum_{j=1}^{n} \dot{g}_j(0) C_j e^{-\frac{\sigma}{2}t}} \right] \end{aligned}$$

*for all* $t \geq 0$ *and* $(i = 1, \cdots, n)$*, where*

$$\sigma = \min_{1\leq i\leq n} \left\{ \frac{1}{C_i R_i}, -\lambda_{max}(\Pi) \cdot \min_{1\leq i\leq n} \{\dot{g}_i(0)\} \right\} > 0. \tag{2.9}$$

*Proof*: By condition (2.8) and (2.9), clearly

$$\left[ -\frac{1 - \sigma R_i C_i}{\dot{g}_i(0)R_i}\delta_{ij} + \frac{T_{ij} + T_{ji}}{2} \right]_{n\times n} \leq 0.$$

Let $u^* = (u_1^*, \cdots, u_n)^T$ be an equilibrium of (2.1). Define a Lyapunov function

$$V(t) = \sum_{i=1}^{n} C_i \int_{u_i^*}^{u_i(t)} [g_i(s) - g_i(u_i^*)]\, ds \tag{2.10}$$

for $t \geq 0$. Clearly, $V(t) \geq 0$. Using (2.10), (2.1) and Lemma 2.2, and computing the derivative of $V(t)$, it follows that

$$\begin{aligned} \dot{V}(t) \quad &= \quad -\sum_{i=1}^{n} \frac{1}{R_i} [g_i(u_i(t)) - g_i(u_i^*)] (u_i(t) - u_i^*) \\ &\qquad + \sum_{i=1}^{n} \sum_{j=1}^{n} T_{ij} [g_i(u_i(t)) - g_i(u_i^*)] \left[ g_j(u_j(t)) - g_j(u_j^*) \right] \\ &= \quad -\sigma \sum_{i=1}^{n} C_i [g_i(u_i(t)) - g_i(u_i^*)] (u_i(t) - u_i^*) \end{aligned}$$

$$
\begin{aligned}
& -\sum_{i=1}^{n}\left(\frac{1}{R_i}-\sigma C_i\right)\left[g_i(u_i(t))-g_i(u_i^*)\right](u_i(t)-u_i^*) \\
& +\sum_{i=1}^{n}\sum_{j=1}^{n}\frac{T_{ij}+T_{ji}}{2}\left[g_i(u_i(t))-g_i(u_i^*)\right]\cdot\left[g_j(u_j(t))-g_j(u_j^*)\right] \\
\leq\ & -\sigma\sum_{i=1}^{n}C_i\left[g_i(u_i(t))-g_i(u_i^*)\right](u_i(t)-u_i^*) \\
& +\sum_{i=1}^{n}\sum_{j=1}^{n}\left[-\frac{1-\sigma R_iC_i}{k_iR_iC_i}\delta_{ij}+\frac{T_{ij}+T_{ji}}{2}\right]\left[g_i(u_i(t))-g_i(u_i^*)\right] \\
& \times\left[g_j(u_j(t))-g_j(u_j^*)\right] \\
\leq\ & -\sigma\sum_{i=1}^{n}C_i\left[g_i(u_i(t))-g_i(u_i^*)\right](u_i(t)-u_i^*) \\
\leq\ & -\sigma\sum_{i=1}^{n}C_i\int_{u_i^*}^{u_i(t)}\left[g_i(s)-g_i(u_i^*)\right]ds \\
=\ & -\sigma V(t)
\end{aligned}
$$

for $t \geq 0$. Then,

$$
V(t) \leq V(0)e^{-\sigma t}, \tag{2.11}
$$

for $t \geq 0$.

By (2.10),

$$
\begin{aligned}
V(0) &= \sum_{i=1}^{n}C_i\int_{u_i^*}^{u_i(0)}\left[g_i(s)-g_i(u_i^*)\right]ds \\
&\leq \sum_{i=1}^{n}C_i\left[g_i(u_i(0))-g_i(u_i^*)\right](u_i(0)-u_i^*) \\
&\leq \sum_{i=1}^{n}\dot{g}_i(0)C_i\left(u_i(0)-u_i^*\right)^2 \\
&\leq \|u(0)-u^*\|^2\sum_{i=1}^{n}\dot{g}_i(0)C_i.
\end{aligned}
$$

Using Lemma 2.2

$$
V(t) \geq C_i\int_{u_i^*}^{u_i(t)}\left[g_i(s)-g_i(u_i^*)\right]ds \geq \frac{C_i}{2\dot{g}_i(0)}\left[g_i(u_i(t))-g_i(u_i^*)\right],
$$

for $t \geq 0$ and $(i = 1, \cdots, n)$. Then, it follows from (2.11) that

$$|g_i(u_i(t)) - g_i(u_i^*)| \leq \|u(0) - u^*\| \sqrt{\frac{2\dot{g}_i(0)}{C_i} \sum_{j=1}^{n} \dot{g}_j(0) C_j e^{-\frac{\sigma}{2}t}} \tag{2.12}$$

for $t \geq 0$ and $(i = 1, \cdots, n)$.

From (2.1) and (2.12), it follows that

$$\begin{aligned} C_i \cdot D^+|u_i(t) - u_i^*| &\leq -\frac{1}{R_i}|u_i(t) - u_i^*| + \sum_{j=1}^{n} |T_{ij}||g_i(u_i(t)) - g_i(u_i^*)| \\ &\leq -\frac{1}{R_i}|u_i(t) - u_i^*| + \|u(0) - u^*\| \sum_{j=1}^{n} |T_{ij}| \\ &\quad \times \sqrt{\frac{2\dot{g}_i(0)}{C_i} \sum_{j=1}^{n} \dot{g}_j(0) C_j e^{-(\sigma/2)t}} \end{aligned}$$

for $t \geq 0$. Integrating throughout, it follows that

$$\begin{aligned} |u_i(t) - u_i^*| &\leq \|u(0) - u^*\| \left[ e^{-\frac{t}{C_i R_i}} + \frac{2R_i}{2 - \sigma R_i C_i} \sum_{j=1}^{n} |T_{ij}| \right. \\ &\quad \left. \times \sqrt{\frac{2\dot{g}_i(0)}{C_i} \sum_{j=1}^{n} \dot{g}_j(0) C_j e^{-(\sigma/2)t}} \right] \end{aligned}$$

for $t \geq 0$. This completes the proof.

Next, we introduce additional parameters to (2.1) enable the development of more flexible results. Suppose $\gamma_i > 0 (i = 1, \cdots, n)$ are some constants. Rewrite (2.1) as

$$\gamma_i C_i \frac{du_i(t)}{dt} = -\frac{\gamma_i}{R_i} u_i(t) + \sum_{j=1}^{n} \gamma_i T_{ij} g_j(u_j(t)) + \gamma_i I_i, (i = 1, \cdots, n)$$

for $t \geq 0$. Then, directly using Theorem (2.7), Theorem 2.9 follows.

THEOREM 2.9 *If there exist constants $\gamma_i > 0 (i = 1, \cdots, n)$ such that*

$$\Pi_\gamma = \left[ -\frac{\gamma_i}{\dot{g}_i(0) R_i C_i} \delta_{ij} + \frac{\gamma_i T_{ij} + \gamma_j T_{ji}}{2} \right]_{n \times n} < 0,$$

*the network (2.1) is globally exponentially stable. Moreover, the network has a unique equilibrium $u^* = (u_1^*, \cdots, u_n^*)^T \in R^n$, and given any $u(0) \in R^n$, the* *trajectory* $u(t) = (u_1(t), \cdots, u_n(t))^T$ *of (2.1) starting from* $u(0)$ *satisfies*

$$\begin{aligned} |u_i(t) - u_i^*| \quad \leq \quad & \|u(0) - u^*\| \left[ e^{-\frac{t}{R_i C_i}} + \frac{2R_i}{2 - \sigma R_i C_i} \right. \\ & \left. \times \sum_{j=1}^{n} |T_{ij}| \sqrt{\frac{2\dot{g}_i(0)}{\gamma_i C_i} \sum_{j=1}^{n} \dot{g}_j(0) \gamma_j C_j e^{-\frac{\sigma t}{2}}} \right] \end{aligned}$$

*for all* $t \geq 0$ *and* $(i = 1, \cdots, n)$, *where*

$$\sigma = \min_{1 \leq i \leq n} \left\{ \frac{1}{C_i R_i}, -\lambda_{max}(\Pi_\gamma) \cdot \min_{1 \leq i \leq n} \{\dot{g}_i(0)\} \right\} > 0.$$

Theorem 2.7 and Theorem 2.8 are rather interesting. They not only give conditions for global exponential stability, but also provide an estimate of the convergence rate of the network (2.1).

## 5. Local Exponential Convergence

It is well known that in the application of a neural network to associative memory, the neural network should have more than one equilibrium. By local convergence of a network, it allows the network to have multi-stable equilibrium points. In this section, some qualitative results on local exponential convergence will be developed. Bounds of the convergence rate and the attractive domain will be given.

DEFINITION 2.1 *An equilibrium* $u^* = (u_1^*, \cdots, u_n^*)^T$ *is called locally exponentially stable, if there exists a* $\lambda > 0$, *and for any* $\epsilon > 0$, *there exists a* $\delta > 0$ *such that* $\|u(0) - u^*\| < \delta$ *implies that*

$$\|u(t) - u^*\| \leq \epsilon e^{-\lambda t},$$

*for all* $t \geq 0$.

LEMMA 2.4 *Let* $D \subset R^n$ *be a convex subset and* $u^* = (u_1^*, \cdots, u_n^*)^T \in D$ *be a fixed point. If there exist constants* $k_i > 0 (i = 1, \cdots, n)$ *such that*

$$0 < \dot{g}_i(u_i) \leq k_i, (i = 1, \cdots, n),$$

*for each* $(u_1, \cdots, u_n)^T \in D$, *then*

$$\int_{u_i^*}^{u_i} [g_i(s) - g_i(u_i^*)] \, ds \geq \frac{1}{2k_i} [g_i(u_i) - g_i(u_i^*)]^2, (i = 1, \cdots, n),$$

*for all* $(u_1, \cdots u_n)^T \in D$.

The proof of Lemma 2.4 is similar to the proof of Lemma 2.3. Details are omitted.

THEOREM 2.10 *Let $(u_1^*, \cdots, u_n^*)^T$ be an equilibrium of (2.1) in the following closed set*

$$D = \{u||u_i - u_i^*| \le \alpha, (i = 1, \cdots, n)\}$$

*where $\alpha > 0$ is a constant. If*

$$\Pi = \left[ -\frac{1}{k_i R_i C_i}\delta_{ij} + \frac{T_{ij} + T_{ji}}{2} \right]_{n \times n} < 0,$$

*the solution $u(t) = (u_1(t), \cdots, u_n(t))^T$ starting from any point*

$$u(0) = (u_1(0), \cdots, u_n(0)) \in Q,$$

*where*

$$Q = \left\{ u||u_i - u_i^*| < \frac{\alpha}{\beta}, (i = 1, \cdots, n) \right\} \subset D$$

*and*

$$\beta = \max_{1 \le i \le n} \left( 1 + \frac{2R_i}{2 - \sigma R_i C_i} \sum_{j=1}^n |T_{ij}| \sqrt{\frac{2k_i}{C_i} \sum_{j=1}^n k_j C_j} \right),$$

*will stay in D and satisfies*

$$\begin{aligned} |u_i(t) - u_i^*| \le\ & \|u(0) - u^*\| \left[ e^{-\frac{t}{R_i C_i}} + \frac{2R_i}{2 - \sigma R_i C_i} \right. \\ & \left. \times \sum_{j=1}^n |T_{ij}| \sqrt{\frac{2k_i}{C_i} \sum_{j=1}^n k_j C_j e^{-\frac{\sigma}{2}t}} \right] \end{aligned}$$

*for all $t \ge 0$ and $(i = 1, \cdots, n)$, where*

$$\sigma = \min_{1 \le i \le n} \left\{ \frac{1}{C_i R_i}, -\lambda_{max}(\Omega) \cdot \min_{1 \le i \le n} \{k_i\} \right\} > 0.$$

*Proof*: Firstly, it will be proven that a trajectory $u(t) = (u_1(t), \cdots, u_n(t))^T$ starting from any initial point $u(0) = (u_1(0), \cdots, u_n(0))^T \in Q$ will stay in $D$ for all $t \ge 0$ and never across the boundary of $D$. Otherwise, since $u(0) \in Q$, there must exist a $t_1 > 0$ and some $i$ such that

$$\begin{cases} |u_i(t_1) - u_i^*| = \alpha \\ |u_j(t) - u_j^*| \le \alpha, \quad t \in [0, t_1], \quad (j = 1, \cdots, n). \end{cases}$$

Define a function

$$V(t) = \sum_{j=1}^n C_i \int_{u_i^*}^{u_i(t)} [g_i(s) - g_i(u_i^*)]\, ds, \quad t \in [0, t_1].$$

Clearly, $V(t) \geq 0$. Then, using Lemma 2.4, similar to the proof of Theorem (2.8), it follows that

$$
\begin{aligned}
|u_i(t) - u_i^*| &\leq \|u(0) - u^*\| \left[ e^{-\frac{t}{C_i R_i}} + \frac{2R_i}{2 - \sigma R_i C_i} \right. \\
&\quad \left. \times \sum_{j=1}^{n} |T_{ij}| \sqrt{\frac{2k_i}{C_i} \sum_{j=1}^{n} k_j C_j e^{-\frac{\sigma}{2} t}} \right]
\end{aligned}
$$

for $t \in [0, t_1]$. Thus, it follows that

$$
\begin{aligned}
|u_i(t_1) - u_i^*| &\leq \|u(0) - u^*\| \left[ e^{-\frac{t_1}{C_i R_i}} + \frac{2R_i}{2 - \sigma R_i C_i} \right. \\
&\quad \left. \times \sum_{j=1}^{n} |T_{ij}| \sqrt{\frac{2k_i}{C_i} \sum_{j=1}^{n} k_j C_j e^{-\frac{\sigma}{2} t_1}} \right] \\
&\leq \|u(0) - u^*\| \cdot \beta \\
&< \alpha.
\end{aligned}
$$

This is contradiction and shows that $x(t) \in D$ for all $t \geq 0$ and never across the boundary of $D$. Using this fact and again follow that proof of Theorem 2.7 the result follows. The proof is completed.

COROLLARY 2.2 *Suppose $x^* = (x_1^*, \cdots, x_n^*)^T$ is an equilibrium of (2.1), if*

$$
\left[ -\frac{1}{\dot{g}_i(x_i^*) \cdot R_i} \delta_{ij} + \frac{T_{ij} + T_{ji}}{2} \right]_{n \times n} < 0,
$$

*$x^*$ is locally exponentially stable.*

## 5.1 Example

EXAMPLE 2.1 *Consider the following Hopfield neural network*

$$
\begin{cases}
\dot{u}_1 = -u_1 + \frac{17 \ln 4}{15} g(u_2) \\
\dot{u}_2 = -u_2 + \frac{17 \ln 4}{15} g(u_1)
\end{cases}
\tag{2.13}
$$

*where $g(s) = \tanh(s)$.*

Clearly, $\dot{g}(s) = 1 - g(s)^2$. It is easy to see that (2.13) has three equilibrium points $(0, 0)$, $(\ln 4, \ln 4)$ and $(-\ln 4, -\ln 4)$. By Corollary 2.2, the equilibria $(\ln 4, \ln 4)$ and $(-\ln 4, -\ln 4)$ are locally exponentially stable. Next, an exponential convergent rate and a domain of attraction will be calculated.

Let

$$D_1 = \left\{ (u_1, u_2)^T \middle| \, |u_i - \ln 4| \le \ln 4(3 - 2\sqrt{2}) \cdot (3 + 2\sqrt{2}), (i = 1, 2) \right\}$$

and

$$D_2 = \left\{ (u_1, u_2)^T \middle| \, |u_i + \ln 4| \le \ln 4(3 - 2\sqrt{2}) \cdot (3 + 2\sqrt{2}), (i = 1, 2) \right\}.$$

Clearly, $(\ln 4, \ln 4) \in D_1$ and $(-\ln 4, -\ln 4) \in D_2$. Since $\dot{g}(s) \le 0.5$ for $|s| \le \ln \sqrt{3 + 2\sqrt{2}}$, then

$$\frac{dg(u_1)}{du_1} \le 0.5, \quad \frac{dg(u_2)}{du_2} \le 0.5$$

for all $(u_1, u_2) \in D_1$ or $(u_1, u_2) \in D_2$. Taking $k = 0.5$, and

$$\Gamma = \begin{bmatrix} -2 & \frac{17 \ln 4}{15} \\ \frac{17 \ln 4}{15} & -2 \end{bmatrix}$$

then,

$$\sigma = \min \{1, -\lambda_{max}(\Pi)\} = 1 - \frac{17 \ln 4}{30}.$$

From Theorem 2.10, it follows that

$$\begin{aligned} Q_1 &= \left\{ \begin{pmatrix} u_1 \\ u_2 \end{pmatrix} \middle| \, |u_i - \ln 4| < \frac{30 + 17 \ln 4}{30 + 85 \ln 4} \right. \\ &\quad \left. \times \ln 4(3 - 2\sqrt{2})(3 + 2\sqrt{2}), (i = 1, 2) \right\} \subset D_1. \end{aligned}$$

Then, any trajectory starting from $(x_1(0), x_2(0)) \in Q$ satisfies

$$\begin{aligned} |u_i(t) - \ln 4| &\le \max \{|u_1(0) - \ln 4|, |u_2(0) - \ln 4|\} \\ &\quad \times \left( e^{-t} + \frac{68 \ln 4}{30 + 17 \ln 4} e^{-\frac{\sigma}{2} t} \right), (i = 1, 2) \end{aligned}$$

for all $t \ge 0$.

Similarly, for the equilibrium $(-\ln 4, -\ln 4) \in D_2$,

$$\begin{aligned} Q_2 &= \left\{ \begin{pmatrix} u_1 \\ u_2 \end{pmatrix} \middle| \, |u_i + \ln 4| < \frac{30 + 17 \ln 4}{30 + 85 \ln 4} \right. \\ &\quad \left. \times \ln 4(3 - 2\sqrt{2})(3 + 2\sqrt{2}), (i = 1, 2) \right\} \subset D_2, \end{aligned}$$

and for any trajectory starting from $(u_1(0), u_2(0))^T \in Q_2$, it satisfies

$$
\begin{aligned}
|u_i(t) + \ln 4| \quad &\leq \quad \max\left\{|u_1(0) - \ln 4|, |u_2(0) - \ln 4|\right\} \\
&\quad \times \left(e^{-t} + \frac{68 \ln 4}{30 + 17 \ln 4} e^{-\frac{g}{2}t}\right), (i = 1, 2)
\end{aligned}
$$

for all $t \geq 0$.

## 6. Discussions and Conclusions

In this chapter, the convergence properties of Hopfield RNNs are studied. Complete stability, global stability and local stability are discussed. Hopfield RNNs have been studied by many researchers and numerous results have been reported on theory and applications. For more results on applications of Hopfield RNNs, interested readers may refer to [64, 4, 190, 145, 101, 172, 109, 105, 63, 202, 31, 187, 103]. For other dynamical properties of RNNs, [201, 155, 21, 49, 151] contain interesting results. For discrete Hopfield RNNs, the readers may refer to [133, 134, 135]. Various conditions for global stability are reported in [57, 60, 110, 215, 104, 74, 96, 31, 1, 200, 209, 35, 125].

Although many stability conditions have been obtained, these conditions are still restrictive. To show this point, consider, as an example, the following simple two dimensional Hopfield RNN

$$
\begin{cases}
\dot{u}_1 = -u_1 + 10 \cdot g(u_2) \\
\dot{u}_2 = -u_2,
\end{cases}
\tag{2.14}
$$

where $g(s) = \tanh(s)$. Clearly, this network is globally convergent, i.e., all trajectories converge to zero point, see Figure 2.2.

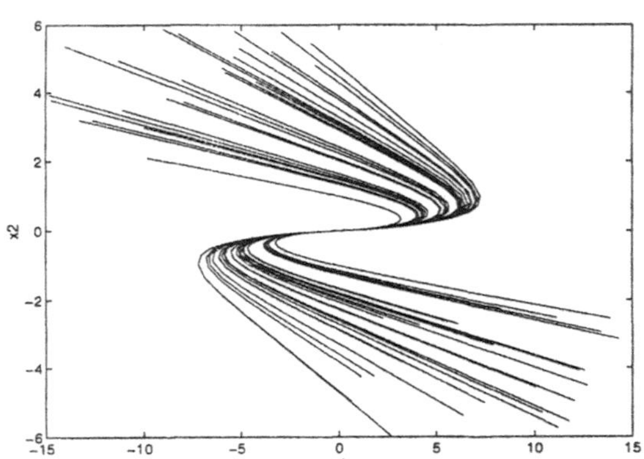


*Figure 2.3.* Global stability of (2.14).

However, the stability of this network cannot be verified via the existing stability conditions reported in [57, 60, 110, 215, 104, 74, 96, 31, 1, 200, 209,

35, 125, 220]. Seeking weaker stability conditions for Hopfield RNNs will continue to pose an interesting challenge.

# Chapter 3

# CELLULAR NEURAL NETWORKS

## 1. Introduciton

The model of Cellular Neural Networks (CNNs) was first proposed by Chua and Yang [42, 43] in 1988. Since then, CNNs have been widely studied both in theory and applications. They are inherently local in nature and are easily to be implemented in very large scale integration (VLSI).

This chapter studies the convergence of CNNs. One of the important characters of CNNs is that their output functions are defined by piecewise linear functions. The convergence of CNNs is strongly related to their output functions. Moreover, the output functions are crucial for CNNs' applications. In Section 2, some important properties of the output functions of CNNs will be discussed.

Three types of CNNs are discussed in this chapter. The first type is standard CNNs. Equilibria, complete stability, global exponential stability as well as estimation of global exponential convergent rate for standard CNNs are analysed in Section 3. Standard CNNs have been successfully applied to signal processing, pattern recognition, associative memories [71, 72], communication problems [53], Euclidean distance transformation [216], especially in processing static images, see for examples, [40, 97, 228, 229, 176, 46, 165, 106].

The second type is CNNs with constant delays. To process moving images, delay parameters should be introduced to be cooperated into CNNs. This motivated the propose of delayed CNNs, see [158, 159]. In Section 4, by constructing novel Lyapunov functionals which contain special terms, global exponential stability conditions will be established.

CNNs with constant delays are simple and ideal cases. In most situations, delays presented in networks are variable and even unbounded. In [218], a model of CNNs with infinite delays is proposed. This model contains both variable and unbounded delays and thus is more general. In Section 5, convergence of CNNs with infinite delays will be studied. Conditions which contains self inhibitions of the networks are established for global convergence.

Finally in Section 6, conclusions follow to summarize this chapter.

## 2. Properties of Output function

The output function $f$ of CNNs is defined as follows:

$$f(s) = \frac{|s+1| - |s-1|}{2}, \quad s \in R. \tag{3.1}$$

The output function $f(\cdot)$ is a piecewise linear function, which is continuous but non-differentiable. Figure 3.1 shows this function. The output function

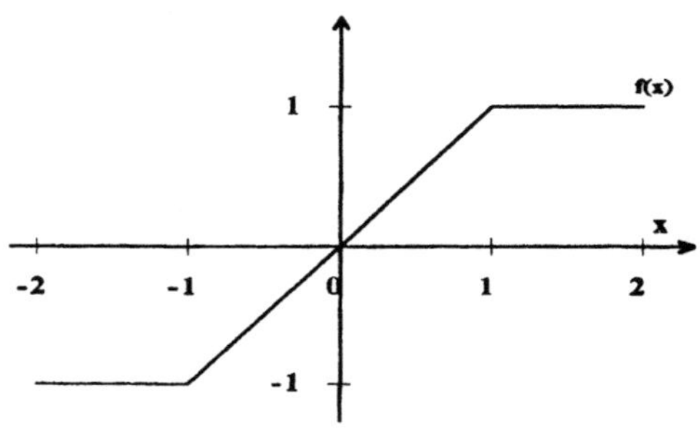


*Figure 3.1.* Output function $f$ of CNNs

$f$ is an important characteristic of CNNs. In fact, many dynamic properties and applications of CNNs are crucially dependent on its properties. In order to study the convergence of CNNs, it is necessary to study the properties of $f$ first. In this section, some useful properties of this function are given.

LEMMA 3.1 *It holds that*

$$(s-\theta)(f(s)-f(\theta)) \geq (f(s)-f(\theta))^2,$$

*for all* $s, \theta \in R$.

LEMMA 3.2 *It holds that*

$$\int_v^u (f(s)-f(v))\, ds \leq (u-v)(f(u)-f(v)),$$

*for all* $u, v \in R$.

LEMMA 3.3 *It holds that*

$$|f(\theta)-f(s)| \leq |\theta - s|,$$

*for all* $\theta, s \in R$.

LEMMA 3.4 *It holds that*

$$sign(\theta - s) \times (f(\theta) - f(s)) = |f(\theta) - f(s)|,$$

for all $\theta, s \in R$,

The proofs of the above lemmas follow directly from the definition of the output function $f(\cdot)$.

LEMMA 3.5 *It holds that*

$$\int_v^u [f(s) - f(v)]\, ds \geq \frac{[f(u) - f(v)]^2}{2},$$

*for all* $u, v \in R$.

*Proof*: Seven cases will be considered to complete the proof.

**Case 1**: $u, v \geq 1$, or $u, v \leq 1$, or $|u|, |v| < 1$. Clearly, the inequality of this lemma is true.

**Case 2**: $u \geq 1$ and $v \leq -1$.

$$\begin{aligned}\int_v^u [f(s) - f(v)]\, ds &= \int_v^{-1} [-1 - (-1)]\, ds \\ &+ \int_{-1}^{1} [s - (-1)]\, ds + \int_1^u [1 - (-1)]\, ds \\ &= 2u \\ &\geq 2 \\ &= \frac{[f(u) - f(v)]^2}{2}.\end{aligned}$$

**Case 3**: $u \leq -1$ and $v \geq 1$.

$$\begin{aligned}\int_v^u [f(s) - f(v)]\, ds &= \int_v^{1} (1 - 1) ds \\ &+ \int_{1}^{-1} (s - 1) ds + \int_{-1}^u (-1 - 1) ds \\ &= -2u \\ &\geq 2 \\ &= \frac{[f(u) - f(v)]^2}{2}.\end{aligned}$$

**Case 4**: $|u| < 1$ and $v \leq -1$.

$$\int_v^u [f(s) - f(v)]\, ds = \int_{-1}^u (s + 1) ds$$

$$
\begin{aligned}
&= \frac{(u+1)^2}{2} \\
&= \frac{[f(u)-f(v)]^2}{2}.
\end{aligned}
$$

**Case 5**: $u \leq -1$ and $|v| < 1$.

$$
\begin{aligned}
\int_v^u [f(s)-f(v)]\,ds &= \int_v^{-1}(s-v)ds + \int_{-1}^u(-1-v)ds \\
&= \frac{(1+v)^2}{2} - (1+v)(u+1) \\
&\geq \frac{(1+v)^2}{2} \\
&= \frac{[f(u)-f(v)]^2}{2}.
\end{aligned}
$$

**Case 6**: $|u| < 1$ and $v \geq 1$.

$$
\begin{aligned}
\int_v^u [f(s)-f(v)]\,ds &= \int_v^{1}(1-1)ds + \int_{1}^u(s-1)ds \\
&= \frac{(u-1)^2}{2} \\
&= \frac{[f(u)-f(v)]^2}{2}.
\end{aligned}
$$

**Case 7**: $u \geq 1$ and $|v| < 1$.

$$
\begin{aligned}
\int_v^u [f(s)-f(v)]\,ds &= \int_v^{1}(s-v)ds + \int_{1}^u(1-v)ds \\
&= \frac{(1-v)^2}{2} + (1-v)(u-1) \\
&\geq \frac{(1-v)^2}{2} \\
&= \frac{[f(u)-f(v)]^2}{2}.
\end{aligned}
$$

The proof is completed.

LEMMA 3.6 *If both $\theta$ and $s$ are in the same interval among $[-1, 1]$ or $(-\infty, -1]$ or $[1, +\infty)$, then*

$$|\theta - c| \leq |s - c|,$$

*implies that*

$$|f(\theta) - f(c)| \leq |f(s) - f(c)|,$$

*for all* $c \in R$.

*Proof*: If $\theta$ and $s$ are both in $(-\infty, -1]$ or both in $[1, +\infty)$, then $f(\theta) = f(s)$. In these cases, the result follows easily.

Next, suppose that both $\theta$ and $s$ are in $[-1, 1]$, then $f(\theta) = \theta$ and $f(s) = s$. Consider three cases:

**Case 1:** If $c \in [-1, 1]$, then $f(c) = c$, and so

$$|f(\theta) - f(c)| = |\theta - c| \leq |s - c| = |f(s) - f(c)|.$$

**Case 2:** If $c \in (-\infty, -1]$, then $f(c) = -1$, and

$$\begin{aligned} |f(\theta) - f(c)| &= \theta + 1 \\ &= |\theta - c| + c + 1 \\ &\leq |s - c| + c + 1 \\ &= s + 1 \\ &= |f(s) - f(c)|. \end{aligned}$$

**Case 3:** If $c \in [1, +\infty)$, then $f(c) = 1$, and

$$\begin{aligned} |f(\theta) - f(c)| &= -\theta + 1 \\ &= |\theta - c| - c + 1 \\ &\leq |s - c| - c + 1 \\ &= |s - 1| \\ &= |f(s) - f(c)|. \end{aligned}$$

The proof is completed.

# 3. Standard CNNs

## 3.1 The Standard CNN Model

The standard CNN model can be described by the following nonlinear differential equation

$$\dot{x}(t) = -x(t) + Af(x(t)) + h, \tag{3.2}$$

or the equivalent form

$$\dot{x}_i(t) = -x_i(t) + \sum_{j=1}^{n} a_{ij} f(x_j) + h_i, \quad (i = 1, \cdots, n),$$

for all $t \geq 0$, where $x_i$ denotes the activity of neuron $i$. $A = (a_{ij})_{n \times n}$ is a real $n \times n$ matrix, each element $a_{ij}$ denotes the strength of the synaptic connection from neuron $j$ onto neuron $i$ and is called synaptic weights. The weights can be positive or negative, corresponding to an excitatory or inhibitory synapse, respectively. $x = (x_1, \cdots, x_n)^T \in R^n$ denotes the state of the network. $h = (h_1, \cdots, h_n)^T \in R^n$ denotes external input. $f(x) = (f(x_1), \cdots, f(x_n))^T$ denotes the output of the network.

## 3.2 Equilibria Analysis

A point $x^* = (x_1^*, \cdots, x_n^*)^T \in R^n$ is called an *equilibrium* of (3.2) if it satisfies

$$-x^* + A \cdot f(x^*) + h \equiv 0, \tag{3.3}$$

or its component form

$$-x_i^* + \sum_{j=1}^{n} a_{ij} f(x_j^*) + h_i \equiv 0$$

for $i = 1, \cdots, n$.

THEOREM 3.1 *The network of (3.2) always has equilibria.*

*Proof:* Define a mapping $\varphi : R^n \to R^n$ as

$$\varphi(x_i) = \sum_{j=1}^{n} a_{ij} f(x_j) + h_i$$

for $i = 1, \cdots, n$. Since $|f(\cdot)| \leq 1$, then

$$|\varphi(x_i)| \leq \sum_{j=1}^{n} |a_{ij}| + |h_i|$$

for $i = 1, \cdots, n$. Obviously, $\varphi$ maps any bounded closed set into a bounded closed set. Then, by the Brouwer fixed point theorem [83], the map $\varphi$ has at least one fixed point $x^* = (x_1^*, \cdots, x_i^*)^T \in R^n$, with $\varphi(x^*) = x^*$, *i.e.*,

$$x_i^* = \sum_{j=1}^{n} a_{ij} f(x_j^*) + h_i, (i = 1, \cdots, n).$$

giving an equilibrium, as required.

THEOREM 3.2 *If there exists a diagonal matrix $D > 0$ such that*

$$-D + \frac{A^T D + DA}{2} < 0$$

*then, the network (3.2) has a unique equilibrium point.*

*Proof*: It is known from Theorem 3.1 that (3.2) has at least one equilibrium. Thus, to complete the proof, it is sufficient to prove uniqueness.

Suppose there are two equilibrium points $x^*$ and $x^{\perp}$. Then, from (3.3),

$$x^* - x^{\perp} = A \cdot \left[f(x^*) - f(x^{\perp})\right]. \tag{3.4}$$

By condition of the theorem, there must exist a sufficiently small constant $\sigma > 0$ such that

$$\begin{cases} D - \sigma I > 0 \\ -D + \sigma I + \dfrac{A^T D + DA}{2} \leq 0, \end{cases}$$

where $I$ is the $n \times n$ identity matrix. Then, from (3.4), it follows that

$$\begin{aligned} 0 &\leq \sigma \cdot \left(x^* - x^{\perp}\right)^T \left[f(x^*) - f(x^{\perp})\right] \\ &= -\left(x^* - x^{\perp}\right)^T (D - \sigma I) \left[f(x^*) - f(x^{\perp})\right] \\ &\quad + \left(x^* - x^{\perp}\right)^T D \left[f(x^*) - f(x^{\perp})\right] \\ &= -\left(x^* - x^{\perp}\right)^T (D - \sigma I) \left[f(x^*) - f(x^{\perp})\right] \\ &\quad + \left[f(x^*) - f(x^{\perp})\right]^T AD \left[f(x^*) - f(x^{\perp})\right]. \end{aligned}$$

Since $D$ is a diagonal matrix, then using Lemma 3.1, it follows that

$$\begin{aligned} &\left(x^* - x^{\perp}\right)^T (D - \sigma I) \left[f(x^*) - f(x^{\perp})\right] \\ \geq \; &\left[f(x^*) - f(x^{\perp})\right]^T (D - \sigma I) \left[f(x^*) - f(x^{\perp})\right]. \end{aligned}$$

Thus,

$$\begin{aligned} 0 &\leq \sigma \left(x^* - x^{\perp}\right)^T \left[f(x^*) - f(x^{\perp})\right] \\ &\leq \left(x^* - x^{\perp}\right)^T \left(-D + \sigma I + \frac{A^T D + DA}{2}\right) \left[f(x^*) - f(x^{\perp})\right] \\ &\leq 0. \end{aligned}$$

That is,

$$\left(x^* - x^{\perp}\right)^T \left[f(x^*) - f(x^{\perp})\right] = 0.$$

Without loss of generality, assume for some $k (1 \leq k \leq n)$ that

$$\begin{cases} x_i^* - x_i^{\perp} = 0, & 1 \leq i \leq k \\ f(x_j^*) - f(x_j^{\perp}) = 0, & k+1 \leq j \leq n. \end{cases}$$

Since $x_i^* - x_i^{\perp} = 0 (i = 1, \cdots, k)$ imply that $f(x_i^*) - f(x_i^{\perp}) = 0 (i = 1, \cdots, k)$, then

$$f(x^*) = f(x^{\perp}).$$

From (3.4), it follows that $x^* = x^{\perp}$. This completes the proof.

## 3.3 Complete Stability of CNNs

The complete stability of a CNN means that every trajectory of the CNN converges to an equilibrium point. This important characteristic of CNNs is critical to many applications. In fact, it represents a fundamental convergent property necessary to apply CNNs to image processing, association memory, etc.

In [42], it is proven that if the connection matrix of a CNN is symmetric, then the CNN will be completely stable. However, as pointed out in [205], the proof is incomplete and it contains some gaps [205]. In [205], a more rigorous proof of complete stability for the symmetric CNN was given by using the LaSalle's invariance principle. However, it requires the activation function to be differentiable with a positive derivative everywhere. This requirement contradicts the piecewise linear activation function used in the original CNN model, which is not differentiable. In [67, 44], the complete stability of the CNNs was also studied, but this aspect of a non-differentiable activation function was also not considered.

To solve this problem, in this section, a novel method is proposed to equivalently represent the trajectories of a CNN via a trajectory to another related differential equation which possesses desirable properties for convergence analysis. Together with a formulation of an appropriate energy-like function, the complete stability of the symmetric CNN will be proven.

THEOREM 3.3 *For any $x_0 \in R^n$, the trajectory $x(t)$ of CNN (3.2) originating from $x_0$ can be represented as*

$$x(t) = (x_0 - h)e^{-t} + Ay(t), \tag{3.5}$$

*where $y(t)$ is the trajectory of the following differential equation*

$$\dot{y} = -y + f\left((x_0 - h)e^{-t} + Ay + h\right), \tag{3.6}$$

*originating from the origin.*

*Proof*: For any $x_0 \in R^n$, let $x(t)$ be the trajectory of CNN (3.2) originating from $x_0$. Consider the following differential equation:

$$\dot{y}(t) = -y(t) + f(x(t)), \tag{3.7}$$

for all $t \geq 0$.

From (3.2) and (3.7), it follows that

$$\frac{d\left[x(t)-h-Ay(t)\right]}{dt}=-\left[x(t)-h-Ay(t)\right],$$

for all $t \geq 0$. Then,

$$x(t)=(x_0-h-Ay(0))e^{-t}+Ay(t)+h,$$

for all $t \geq 0$, where $y(0) \in R^n$ is an initial vector of (3.7). Taking $y(0)=0$ in particular, then $y(t)$ is the trajectory of (3.7) originating from the origin. Moreover,

$$x(t)=(x_0-h)e^{-t}+Ay(t)+h, \tag{3.8}$$

for all $t \geq 0$. The result now follows by substituting (3.8) into (3.7). The proof is completed.

Therefore, following Theorem 3.3, the trajectories of the CNN (3.2) can be represented by a trajectory of (3.6) plus a decaying term. However, it should be noted that the trajectory $y(t)$ depends on the initial value $x_0$ of CNN (3.2).

The equivalent representation of the trajectories of CNN (3.2) with a trajectory of the differential equation (3.6) is rather interesting. From the representation (3.5), it is easy to see that the convergence of the trajectories of (3.2) can be transformed to the convergence of the trajectory of (3.6) starting from the origin. Thus, it provides a novel method to analyse the convergence of CNNs.

Let $\mathcal{E}$ be the set of all equilibrium points of the CNN (3.2).

Denote

$$Q=\left\{x \in R^n \,|-1 \leq x_i \leq 1, (i=1,\cdots,n)\right\}.$$

It is easy to see that $x \in Q$ if and only if $|f(x_i)| = x_i(i=1,\cdots,n)$.

DEFINITION 3.1 *The CNN (3.2) is said to be completely stable, if for any initial vector $x_0 \in R^n$, there exists an equilibrium $x^* \in \mathcal{E}$ such that the trajectory of $x(t,x_0)$ originating from $x_0$ satisfies*

$$\lim_{t\to+\infty} x(t)=x^*.$$

Next, consider the convergence of the trajectory of (3.6) starting from origin. The differential equation (3.6) has certain desirable properties for convergence analysis.

THEOREM 3.4 *For any $x_0 \in R^n$, the trajectory $y(t)$ of (3.6) originating from the origin satisfies*

$$y(t) \in Q,$$

*for all $t \geq 0$.*

*Proof*: Rewrite (3.6) by

$$\dot{y}_i(t) = -y_i(t) + f_i\left((x_0^i - h_i)e^{-s} + \sum_{j=1}^{n} a_{ij}y_j(s) + h_i\right), \tag{3.9}$$

for $t \geq 0$ and $i = 1, \cdots, n$.

Since $y(0) = 0$, then from (3.9), it follows that

$$y_i(t) = \int_0^t e^{-(t-s)} f_i\left((x_0^i - h_i)e^{-s} + \sum_{j=1}^{n} a_{ij}y_j(s) + h_i\right) ds,$$

and so

$$\begin{aligned}
|y_i(t)| &\leq \int_0^t e^{-(t-s)} \left| f_i\left((x_0^i - h_i)e^{-s} + \sum_{j=1}^{n} a_{ij}y_j(s) + h_i\right)\right| ds \\
&\leq \int_0^t e^{-(t-s)} ds \\
&= 1 - e^{-t} \\
&\leq 1,
\end{aligned}$$

for all $t \geq 0$ and $(i = 1, \cdots, n)$. Thus, $y(t) \in Q$ for all $t \geq 0$. This completes the proof.

Next, an energy-like function is constructed to prove the convergence of $y(t)$.

THEOREM 3.5 *Suppose the matrix A is symmetric. Then, for any $x_0 \in R^n$, the trajectory $y(t)$ of (3.5), originating from the origin, will converge to a constant vector $y^*(x_0) \in R^n$ which satisfies*

$$-y^*(x_0) + f\left(Ay^*(x_0) + h\right) = 0.$$

*Proof*: Constructing an energy-like function

$$\begin{aligned}
E(t) &= \frac{1}{2}y^T(t)(I - A)y(t) - h^T y(t) \\
&\quad -(x_0 - h)^T\left(y(t)e^{-t} + \int_0^t y(s)e^{-s}ds\right)
\end{aligned} \tag{3.10}$$

for all $t \geq 0$.

By Theorem 3.4, $y(t) \in Q$ for all $t \geq 0$, then $\|y(t)\| \leq \sqrt{n}$ for all $t \geq 0$. It follows from (3.10) that

$$\begin{aligned}
|E(t)| &\leq \frac{n}{2}\|I - A\| + \sqrt{n}\|h\| \\
&\quad +\|x_0 - h\|\left(\sqrt{n}e^{-t} + \sqrt{n}\int_0^t e^{-s}ds\right) \\
&\leq \frac{n}{2}\|I - A\| + \sqrt{n}\|h\| + \sqrt{n}\|x_0 - h\|,
\end{aligned}$$

for all $t \geq 0$. Thus, $E(t)$ is bounded.

Since $y(t) \in D$ for all $t \geq 0$, $f(y(t)) = y(t)$ for all $t \geq 0$, it follows from (3.10) and (3.6) that

$$\begin{aligned}
\dot{E}(t) &= \left[y(t) - \left((x_0 - h)e^{-t} + Ay(t) + h\right)\right]^T \\
&\quad \times \left[-y(t) + f\left((x_0 - h)e^{-t} + Ay(t) + h\right)\right] \\
&= \left[y(t) - \left((x_0 - h)e^{-t} + Ay(t) + h\right)\right]^T \\
&\quad \times \left[-f(y(t)) + f\left((x_0 - h)e^{-t} + Ay(t) + h\right)\right] \\
&\leq -\|f(y(t)) - f\left((x_0 - h)e^{-t} + Ay(t) + h\right)\|^2 \\
&= -\|y(t) - f\left((x_0 - h)e^{-t} + Ay(t) + h\right)\|^2 \\
&= -\|\dot{y}(t)\|^2,
\end{aligned}$$

for all $t \geq 0$. Thus, $E(t)$ is monotonically decreasing. Since $E(t)$ is bounded, there must exist a constant $E_0$ such that

$$\lim_{t \to +\infty} E(t) = E_0 < -\infty.$$

Noting that $\int_0^t \|\dot{y}(s)\|^2 ds$ is monotonically increasing, it follows that

$$\begin{aligned}
\int_0^{+\infty} \|\dot{y}(s)\|^2 ds &= \lim_{t \to +\infty} \int_0^t \|\dot{y}(s)\|^2 \\
&\leq \lim_{t \to +\infty} (E(0) - E(t)) \\
&= E(0) - E_0 \\
&< +\infty.
\end{aligned}$$

Since $y(t) \in Q$ for all $t \geq 0$, it is clear from (3.6) that $\|\dot{y}(t)\|^2$ is uniformly continuous on $[0, +\infty)$. Thus,

$$\lim_{t \to +\infty} \|\dot{y}(t)\|^2 = 0$$

and

$$\lim_{t \to +\infty} \dot{y}(t) = 0.$$

By the definition of the output function $f$ and using the fact that $A$ is a symmetric matrix, the trajectory $y(t)$ must be a function as

$$y(t) = F\left(e^{\lambda_1 t}, \cdots, e^{\lambda_m t}\right)$$

where $F$ is a polynomial and $\lambda_1, \cdots, \lambda_m$ are constants. Thus, $\dot{y}(t) \to 0$ as $t \to +\infty$ implies that $y(t)$ is convergent, i.e., there exists a constant vector $y^*(x_0) \in R^n$, which depends on $x_0$, such that

$$\lim_{t \to +\infty} y(t) = y^*(x_0).$$

Clearly,

$$-y^*(x_0) + f\left(Ay^*(x_0) + h\right) = 0.$$

This completes the proof.

Next, the complete stability of the CNN (3.2) is proven.

THEOREM 3.6 *Suppose matrix $A$ is symmetry, then the CNN (3.2) is completely stable.*

*Proof*: For any $x_0 \in R^n$, following Theorem 3.1, the trajectory of (3.2) originating from this point can be represented as

$$x(t) = (x_0 - b)e^{-t} + Ay(t) + h,$$

for $t \geq 0$, where $y(t)$, which depends on $x_0$, is the trajectory of (3.6) starting from the origin. By Theorem 3.5, there exists a constant $y^*(x_0)$ such that

$$\lim_{t \to +\infty} y(t) = y^*(x_0),$$

and

$$-y^*(x_0) + f\left(Ay^*(x_0) + h\right) = 0.$$

Then,

$$\lim_{t \to +\infty} x(t) = Ay^*(x_0) + h.$$

Denote

$$x^* = Ay^*(x_0) + h,$$

it follows that

$$-x^* + Af(x^*) + h = A\left[-y^*(x_0) + f\left(Ay^*(x_0) + h\right)\right] = 0.$$

This shows that $x^*$ is an equilibrium of the CNN (3.2). The proof is completed.

Next, a simple extension to the above results, which relaxes the requirement on matrix $A$ to be symmetry is given.

THEOREM 3.7 *If there exists a diagonal matrix $D$ such that the matrix $DA$ is symmetric, then the CNN (3.2) is completely stable.*

*Proof*: Define an energy-like function as follows:

$$\begin{aligned} E(t) &= \frac{1}{2}y^T(t)D(I-A)y(t) - h^T Dy(t) \\ &\quad -(x_0 - h)^T D\left(y(t)e^{-t} + \int_0^t y(s)e^{-s}ds\right), \end{aligned}$$

for all $t \geq 0$. The result now follows along a similar discussion as for Theorem 3.6. The details are thus omitted. The proof is completed.

## 3.4 Global Exponential Stability and Convergence Rate Estimation

Global exponential stability (GES) is a good convergence property of dynamic systems. It can provide a clear understanding of the convergence, which is especially interesting in engineering field. This section will establish GES conditions for the network (3.2). Meanwhile, the GES convergence rate estimation will also be given.

DEFINITION 3.2 *An equilibrium $x^*$ of the CNN (3.2) is called GES, if there exist constants $\epsilon > 0$ and $M \geq 1$ such that*

$$|x_i(t) - x_i^*| \leq M\|x(0) - x^*\|e^{-\epsilon t}$$

*for all $t \geq 0$.*

THEOREM 3.8 *If*

$$\lambda_{max}\left(\frac{A^T + A}{2}\right) < 1,$$

*the CNN (3.2) is GES. Moreover,*

$$|x_i(t) - x_i^*| \leq \|x(0) - x^*\|\left[e^{-t} + \frac{2\sqrt{2}\sum_{j=1}^n \bar{a}_{ij}}{1+\lambda^+}\left(e^{-\frac{1-\lambda^+}{2}t} - e^{-t}\right)\right]$$

*for all $t \geq 0$ and $(i = 1, \cdots, n)$, where*

$$\lambda^+ = \max\left\{0, \lambda_{max}\left(\frac{A^T + A}{2}\right)\right\}, \quad \bar{a}_{ij} = \begin{cases} \max\{0, a_{ii}\}, & i = j \\ |a_{ij}|, & i \neq j. \end{cases}$$

*Proof*: Clearly, conditions of Theorem 3.2 are satisfied. Then, the CNN (3.2) has only one equilibrium $x^*$.

Define a differentiable function

$$V(t) = \sum_{j=1}^n \int_{x_i^*}^{x_i(t)} [f(s) - f(x_i^*)]\, ds$$

for $t \geq 0$. Obviously, $V(t) \geq 0$ for all $t \geq 0$.

Using Lemma 3.1 and Lemma 3.2, it follows that

$$\begin{aligned}
\dot{V}(t) &= -[f(x(t)) - f(x^*)]^T \cdot (x(t) - x^*) \\
&\quad + [f(x(t)) - f(x^*)]^T \left(\frac{A^T + A}{2}\right) [f(x(t)) - f(x^*)] \\
&\leq -[f(x(t)) - f(x^*)]^T \cdot (x(t) - x^*) \\
&\quad + \lambda^+ \cdot [f(x(t)) - f(x^*)]^T [f(x(t)) - f(x^*)] \\
&\leq -(1 - \lambda^+) \cdot [f(x(t)) - f(x^*)]^T \cdot (x(t) - x^*) \\
&\leq -(1 - \lambda^+) \cdot V(t)
\end{aligned}$$

for $t \geq 0$. Then,

$$V(t) \leq V(0) e^{-(1-\lambda^+)t}$$

for $t \geq 0$. By Lemma 3.2,

$$\begin{aligned}
V(0) &= \sum_{i=1}^{n} \int_{x_i^*}^{x_i(0)} [f(s) - f(x_i^*)]\, ds \\
&\leq \sum_{i=1}^{n} [f(x_i(0)) - f(x_i^*)]\,(x_i(0) - x_i^*) \\
&\leq \sum_{i=1}^{n} [x_i(0) - x_i^*]^2 \\
&= \|x(0) - x^*\|^2.
\end{aligned}$$

On the other hand, from Lemma 3.5,

$$V(t) \geq \frac{1}{2} \|f(x(t)) - f(x^*)\|^2$$

for $t \geq 0$. Thus,

$$\|f(x(t)) - f(x^*)\| \leq \sqrt{2} \|x(0) - x^*\| e^{-\frac{1-\lambda^+}{2} t}$$

for $t \geq 0$.

From (3.2),

$$\frac{d\,[x_i(t) - x_i^*]}{dt} = -[x_i(t) - x_i^*] + \sum_{j=1}^{n} a_{ij} \left[f(x_j(t)) - f(x_j^*)\right]$$

for $t \geq 0$ and $(i = 1, \cdots, n)$. Then,

$$D^+ |x_i(t) - x_i^*| = -|x_i(t) - x_i^*| + \sum_{j=1}^{n} \bar{a}_{ij} \left|f(x_j(t)) - f(x_j^*)\right|$$

$$\leq \; -|x_i(t) - x_i^*| + \sqrt{2}\sum_{j=1}^{n} \bar{a}_{ij}\|x(0) - x^*\|e^{-\frac{1-\lambda^+}{2}t}$$

for $t \geq 0$. Thus, it follows that

$$\begin{aligned}|x_i(t) - x_i^*| \;&\leq\; -|x_i(0) - x_i^*|\, e^{-t} \\ &\quad + \sqrt{2}\sum_{j=1}^{n} \bar{a}_{ij}\|x(0) - x^*\| \cdot \int_0^t e^{-(t-s)}e^{-\frac{1-\lambda^+}{2}s}ds \\ &\leq\; \|x(0) - x^*\|\left[e^{-t} + \frac{2\sqrt{2}\sum_{j=1}^{n}\bar{a}_{ij}}{1+\lambda^+}\left(e^{-\frac{1-\lambda^+}{2}t} - e^{-t}\right)\right]\end{aligned}$$

for $t \geq 0$ and $(i = 1, \cdots, n)$. This completes the proof.

The above theorem leads to a simple and interesting corollary.

COROLLARY 3.1 *If*

$$\lambda_{max}\left(\frac{A^T + A}{2}\right) \leq 0,$$

*the CNN (3.2) is GES, and*

$$|x_i(t) - x_i^*| \leq \|x(0) - x^*\|\left[e^{-t} + 2\sqrt{2}\sum_{j=1}^{n}\bar{a}_{ij}\left(e^{-\frac{t}{2}} - e^{-t}\right)\right]$$

*for all* $t \geq 0$ *and* $(i = 1, \cdots, n)$.

Consider the following simple one dimensional CNN

$$\dot{x} = -x + af(x) + h$$

for $t \geq 0$. Suppose $a \leq 0$, by the Corollary 3.1, this network is GES and there exists one equilibrium $x^*$ such that

$$|x(t) - x^*| \leq |x(0) - x^*|e^{-t}$$

for all $t \geq 0$.

Conditions in Theorem 3.8 can be further relaxed by introducing some parameters.

THEOREM 3.9 *If there exists a diagonal matrix* $D = diag(d_1, \cdots, d_n)$ *with* $d_i > 0(i = 1, \cdots, n)$ *such that*

$$-D + \frac{A^T D + DA}{2} < 0$$

*then, the CNN (3.2) is GES. Moreover,*

$$|x_i(t)-x_i^*| \leq \|x(0)-x^*\| \left[ e^{-t} + \frac{2\sqrt{2\max_{1\leq l\leq n}\{d_l\}}}{2-\sigma} \cdot \sum_{j=1}^{n} \frac{\bar{a}_{ij}}{\sqrt{d_i}} \left( e^{-\frac{\sigma}{2}t} - e^{-t} \right) \right]$$

*for all t* $\geq 0$ *and* $(i = 1, \cdots, n)$*, where*

$$\sigma = \max \left\{ 1, -\frac{1}{\max_{1\leq i\leq n}\{d_i\}} \cdot \lambda_{max} \left( -D + \frac{A^T D + DA}{2} \right) \right\}$$

*and*

$$\bar{a}_{ij} = \begin{cases} \max\{0, a_{ii}\}, & i = j \\ |a_{ij}|, & i \neq j. \end{cases}$$

*Proof*: Clearly,

$$\begin{cases} (1-\sigma)D \geq 0 \\ \Omega \triangleq -(1-\sigma)D + \frac{A^T D + DA}{2} \leq 0. \end{cases}$$

Define

$$V(t) = \sum_{j=1}^{n} d_i \int_{x_i^*}^{x_i(t)} [f(s) - f(x_i^*)]\, ds$$

for $t \geq 0$.

Using Lemma 3.1 and Lemma 3.2, it follows that

$$\begin{aligned} \dot{V}(t) &= -[f(x(t)) - f(x^*)]^T D(x(t) - x^*) \\ &\quad + [f(x(t)) - f(x^*)]^T \left( \frac{A^T D + DA}{2} \right) [f(x(t)) - f(x^*)] \\ &= -\sigma \cdot [f(x(t)) - f(x^*)]^T D(x(t) - x^*) \\ &\quad - [f(x(t)) - f(x^*)]^T (1-\sigma) D(x(t) - x^*) \\ &\quad + [f(x(t)) - f(x^*)]^T \left( \frac{A^T D + DA}{2} \right) [f(x(t)) - f(x^*)] \\ &\leq -\sigma V(t) + [f(x(t)) - f(x^*)]^T \Omega [f(x(t)) - f(x^*)] \\ &\leq -\sigma V(t) \end{aligned}$$

for $t \geq 0$. The rest part of the proof can be completed via a slight modification to the proof of Theorem 3.8.

In the above theorems, it is easy to see that both the GES conditions and the estimations of convergence rates are totally independent of the network inputs.

Next, some examples are given to illustrate the above developed theory.

EXAMPLE 3.1 *Consider the network*

$$\begin{cases} \dot{x}_1(t) = -x_1(t) + \frac{1}{4}f(x_1(t)) - f(x_2(t)) + h_1 \\ \dot{x}_2(t) = -x_2(t) + 3f(x_1(t)) - \frac{5}{4}f(x_2(t)) + h_2 \end{cases} \quad (3.11)$$

*for t $\geq 0$, where $h_i(i = 1, 2)$ are any constants.*

It is easy to calculate that

$$\lambda_{max}\left(\frac{A^T + A}{2}\right) = 0.75 = \lambda^+ < 1.$$

By Theorem 3.8,

$$\begin{cases} |x_1(t) - x_1^*| \leq \|x(0) - x^*\| \left[e^{-t} + \frac{10\sqrt{2}}{7}\left(e^{-\frac{t}{8}} - e^{-t}\right)\right] \\ |x_2(t) - x_2^*| \leq \|x(0) - x^*\| \left[e^{-t} + \frac{24\sqrt{2}}{7}\left(e^{-\frac{t}{8}} - e^{-t}\right)\right] \end{cases}$$

for $t \geq 0$.

Figure 3.2 shows the simulation result for GES of the network (3.11) for $h_1 = h_2 = 1$. The 60 trajectories in the figure start from randomly selected initial points.

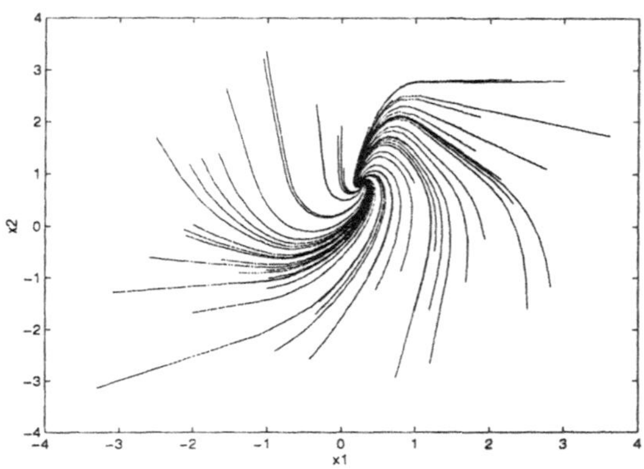


*Figure 3.2.* GES of (3.11) for $h_1 = h_2 = 1$

EXAMPLE 3.2 *Consider a three dimensional network*

$$\dot{x}(t) = -x(t) + Af(x(t)) + h \quad (3.12)$$

*for* $t \geq 0$, *where*

$$A = \begin{bmatrix} -1.0367 & -0.8539 & -0.0287 \\ -0.8539 & -0.7182 & -0.0257 \\ -0.0287 & -0.0257 & -0.0011 \end{bmatrix}, \quad h = \begin{bmatrix} 0.9921 \\ 0.9684 \\ 0.8512 \end{bmatrix}.$$

The matrix $A$ is symmetry. It is easy to check that all the eigenvalues of $A$ are non-positive, i.e., $A \leq 0$. Thus, by Corollary 3.1, there exist a equilibrium $x^*$ such that

$$\begin{cases} |x_1(t) - x_1^*| \leq \|x(0) - x^*\| \left[e^{-t} + 2\sqrt{2} \times 0.8826 \times \left(e^{-\frac{t}{2}} - e^{-t}\right)\right] \\ |x_2(t) - x_2^*| \leq \|x(0) - x^*\| \left[e^{-t} + 2\sqrt{2} \times 0.8796 \times \left(e^{-\frac{t}{2}} - e^{-t}\right)\right] \\ |x_3(t) - x_3^*| \leq \|x(0) - x^*\| \left[e^{-t} + 2\sqrt{2} \times 0.0544 \times \left(e^{-\frac{t}{2}} - e^{-t}\right)\right] \end{cases}$$

for $t \geq 0$.

Figure 3.3 shows the simulation for 80 trajectories starting from randomly selected initial points.

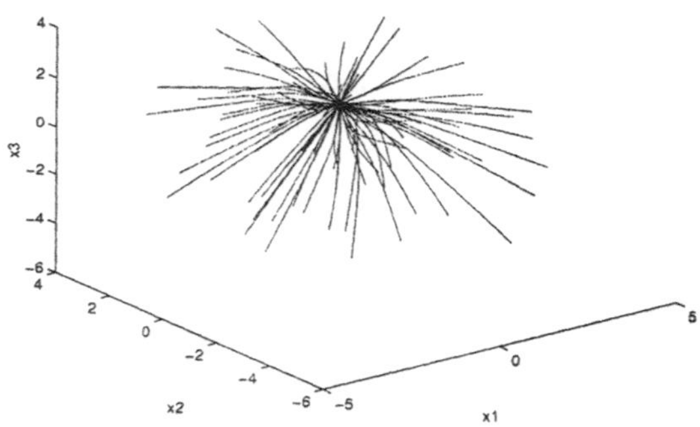


*Figure 3.3.* GES of (3.12)

It can be observed that all the trajectories converge to the point $x^* \approx (0.3085, 0.3977, 0.8312)^T$ very quickly.

## 4. CNNs with Constant Delays

To process moving images, one must introduce delays in the signals transmitted among the cells. This leads to the model of Cellular Neural Networks with delay (DCNNs), introduced in [158, 159]. They have found applications in different areas such as classification of patterns and reconstruction of moving images. Neural networks with delays have attracted wide interests from many authors in recent years, see for example [10, 44, 67, 137, 158, 159, 160, 179, 181, 205, 220, 224, 225].

The convergence properties between CNNs with and without delay are essentially different. How does delay affect the convergence of neural networks? For Hopfield neural networks, as in many other dynamical systems, it is well known that delays may result in instability, see, [17, 137]. They can also affect convergence. In [44], it was shown that delays added to a stable CNN can make it unstable. So, to study CNNs with delays, one must address the problem of how to remove this destabilizing effect.

In this section, the global exponential stability of CNNs with constant delays is studied via constructing novel Lyapunov functionals. Conditions for GES will be established.

## 4.1 Model of CNNs with Constant Delays

The model of DCNNs can be described as following differential difference equation

$$\dot{x}(t) = -x(t) + Af(x(t)) + Bf(x(t-\tau)) + h \tag{3.13}$$

or the equivalent form

$$\dot{x}_i(t) = -x_i(t) + \sum_{j=1}^{n} [a_{ij}f(x_j(t) + b_{ij}f(x_j(t-\tau_{ij}))] + h_i$$

for $t \geq 0$ and $(i = 1, \cdots, n)$, where the delays $\tau_{ij}(i, j = 1, \cdots, n)$ are nonnegative constants.

Let $\tau = \max_{1\leq i,j\leq n}\{\tau_{ij}\}$. The initial conditions for DCNN is defined by

$$x_i(t) = \phi_i(t), t \in [-\tau, 0], (i = 1, \cdots, n)$$

where $\phi_i(t)(i = 1, \cdots, n)$ are continuous functions.

A point $x^* \in R^n$ is an equilibrium point (3.13), if

$$-x^* + (A + B)f(x^*) + h = 0.$$

Denote

$$\|\phi - x^*\| = \max_{1\leq i\leq n}\left[\sup_{-\tau\leq s\leq 0} |\phi_i(s) - x_i^*|\right].$$

DEFINITION 3.3 *The DCNN (3.13) is called globally exponentially stable (GES), if the DCNN (3.13) has only one equilibrium point $x^*$, and there exist constants $\epsilon > 0$ and $M \geq 1$ such that*

$$|x_i(t) - x_i^*| \leq M\|\phi - x^*\|e^{-\epsilon t}, (i = 1, \cdots, n)$$

*for all $t \geq 0$.*

## 4.2 Conditions for GES

THEOREM 3.10 *If there exist constants $\beta_i > 0 (i = 1, \cdots, n)$ such that*

$$-1 + a_{jj} + \frac{1}{\beta_j} \sum_{i=1}^{n} [|a_{ij}|(1 - \delta_{ij}) + |b_{ij}|] < 0, (j = 1, \cdots, n)$$

*where*

$$\delta_{ij} = \begin{cases} 1, & if\ i = j \\ 0, & if\ i \neq j \end{cases}$$

*then, the network (3.13) is GES.*

*Proof*: Since any CNN always has an equilibrium, let $x^*$ be any of the equilibrium points of (3.13). Rewrite (3.13) as

$$\begin{aligned} \frac{d[x_i(t) - x_i^*]}{dt} &= -[x_i(t) - x_i^*] + \sum_{j=1}^{n} a_{ij} \left[ f(x_j(t)) - f(x_j^*) \right] \\ &\quad + \sum_{j=1}^{n} b_{ij} \left[ f(x_j(t - \tau_{ij})) - f(x_j^*) \right] \end{aligned}$$

for $t \geq 0$ and $(i = 1, \cdots, n)$.

Using Lemma 3.3 and Lemma 3.4,

$$\begin{aligned} D^+ |x_i(t) - x_i^*| &\leq -(1 - a_{ii}) |f(x_i(t)) - f(x_i^*)| \\ &\quad + \sum_{j=1, j\neq i}^{n} |a_{ij}| \cdot \left| f(x_j(t)) - f(x_j^*) \right| \\ &\quad + \sum_{j=1}^{n} |b_{ij}| \cdot \left| f(x_j(t - \tau_{ij})) - f(x_j^*) \right| \end{aligned}$$

for $t \geq 0$ and $(i = 1, \cdots, n)$.

By condition of Theorem 3.10, there must exist a small constant $\epsilon > 0$ such that

$$\eta_j \triangleq 1 - a_{jj} - \epsilon - \frac{1}{\beta_j} \sum_{i=1}^{n} \beta_i [|a_{ij}|(1 - \delta_{ij}) + |b_{ij}|] > 0$$

for $(i = 1, \cdots, n)$.

Constructing the following Lyapunov functional

$$\begin{aligned} V(t) &= \sum_{i=1}^{n} \beta_i \Bigg[ |x_i(t) - x_i^*| e^{\epsilon t} \\ &\quad + \sum_{j=1}^{n} |b_{ij}| \int_{t-\tau_{ij}}^{t} \left| f(x_j(s)) - f(x_j^*) \right| e^{\epsilon(s+\tau_{ij})} ds \Bigg] \end{aligned} \quad (3.14)$$

for $t \geq 0$.

It follows that

$$\begin{aligned}
D^+V(t) &= \sum_{i=1}^{n} \beta_i \Bigg[ D^+ \left[ |x_i(t) - x_i^*| \, e^{\epsilon t} \right] \\
&\quad + \sum_{j=1}^{n} |b_{ij}| \left| f(x_j(t)) - f(x_j^*) \right| e^{\epsilon(t+\tau_{ij})} \\
&\quad - \sum_{j=1}^{n} |b_{ij}| \left| f(x_j(t-\tau_{ij})) - f(x_j^*) \right| e^{t} \\
&\leq e^{\epsilon t} \sum_{i=1}^{n} \beta_i \Bigg[ (\epsilon - 1 + a_{ii}) \, |f(x_i(t)) - f(x_i^*)| \\
&\quad \sum_{j=1}^{n} (|a_{ij}|(1-\delta_{ij}) + |b_{ij}|) \left| f(x_j(t)) - f(x_j^*) \right| \Bigg] \\
&\leq -e^{\epsilon t} \sum_{j=1}^{n} \eta_j \left| f(x_j(t)) - f(x_j^*) \right| \\
&\leq 0
\end{aligned}$$

for $t \geq 0$. Then, $V(t) \leq V(0)$ for $t \geq 0$. Since

$$V(t) \geq e^{\epsilon t} \min_{1 \leq j \leq n} \{\beta_j\} \sum_{i=1}^{n} |x_i(t) - x_i^*|$$

for $t \geq 0$, and

$$V(0) \leq \left[ \max_{1 \leq j \leq n} \{\beta_j\} + \tau e^{\epsilon \tau} \sum_{i=1}^{n} \beta_i \max_{1 \leq j \leq n} \{|a_{ij}|\} \right] \|\phi - x^*\|$$

then,

$$\sum_{j=1}^{n} |x_i(t) - x_i^*| \leq M \|\phi - x^*\| e^{-\epsilon t}$$

for $t \geq 0$, where

$$M = \frac{1}{\min_{1 \leq j \leq n} \{\beta_j\}} \left[ \max_{1 \leq j \leq n} \{\beta_j\} + \tau e^{\epsilon \tau} \sum_{i=1}^{n} \beta_i \max_{1 \leq j \leq n} \{|a_{ij}|\} \right].$$

The DCNN (3.13) cannot has another equilibrium which differ from $x^*$. In fact, if there exists another equilibrium $x^\dagger$, it holds that

$$\sum_{j=1}^{n} |x_i^\dagger - x_i^*| \leq M \|x_i^\dagger - x^*\| e^{-\epsilon t}$$

for $t \geq 0$. Clearly, it must hold that $x^{\dagger} = x^*$. Thus, the network (3.13) is GES. The proof is completed.

The Lyapunov functional defined by (3.14) is interesting. It contains a special term $e^{\epsilon t}$ which enables to establish exponential stability condition. The idea of cooperating the special term $e^{\epsilon t}$ into Lyapunov functionals is first proposed in [224].

The conditions in Theorem 3.10 are independent of the delays, and thus are easy to verify.

EXAMPLE 3.3 *Consider the DCNN*

$$\begin{cases} \dot{x}_1(t) = -x_1(t) + \frac{1}{2}f(x_1(t)) + \frac{1}{4}f(x_2(t-\tau_1)) + h_1 \\ \\ \dot{x}_2(t) = -x_2(t) + \frac{3}{2}f(x_1(t-\tau_2)) - f(x_2(t)) + h_2 \end{cases}$$

*where* $\tau_i \geq 0$ *and* $h_i$ *are constants,* $(i = 1, 2)$.

This network satisfies all the conditions of the Theorem 3.10. Thus, it is GES. The delay $\tau_i (i = 1, 2)$ could be any uncertain nonnegative constants.

# 5. CNNs with Infinite Delay

## 5.1 Model of CNNs with Infinite Delay

Most research on DCNNs has been restricted to simple cases of constant delays. Few considers variable or unbounded delay. In CNNs, it is clear that a constant delay is only a special case. In most situations, delays are variable, and even unbounded. That is, the entire history affects the present. Such delay terms, more suitable to practical neural nets, are called ***infinite delays***.

From the mathematical point of view, systems with constant delay are different from those with variable or/and infinite delays, and known mathematical methods do not directly apply. This provides new challenges. In this section, some methods from functional differential equations are developed to study the global convergence of CNNs with infinite delay.

The model of CNNs with infinite delay is described by

$$\begin{aligned} \frac{dx_i(t)}{dt} &= \sum_{j=1}^{n} \left[a_{ij} f(x_j(t)) + b_{ij} f(x_j(t - \tau_{ij}(t)))\right] \\ &\quad + \sum_{j=1}^{n} c_{ij} \int_{-\infty}^{t} k_{ij}(t-s) f(x_j(s)) ds \\ &\quad - x_i(t) + I_i, \quad t \geq 0, (i = 1, \cdots, n) \end{aligned} \tag{3.15}$$

where $a_{ij}, b_{ij}, c_{ij}$ and $I_i$ are constants, $k_{ij} : R^+ \to R^+$ are continuous functions and $\int_0^{+\infty} k_{ij}(s) ds = 1$, $\tau_{ij}(t) (i = 1, 2, \cdots, n)$ are nonnegative continuous functions with $0 \leq \tau_{ij}(t) \leq \tau$, for some constant $\tau$. The $n$-tuple $(x_1(t), \cdots, x_n(t))$ denotes the state of the network at time $t$.

This model contains both variable and infinite delays. Thus, it is more general. Since many motion-related phenomena can be represented and/or modeled by DCNNs [159], this model could be excepted to have applications in processing moving images and associative memories, etc.

Model (3.15) includes many interesting special models. If $c_{ij} = 0(i, j = 1, \cdots, n)$, then (3.15) becomes a model of CNNs with variable delay. Moreover, if the $\tau_{ij}(t)(i, j = 1, \cdots, n)$ are constants, (3.15) reduces to the DCNN model that has been widely studied. If $c_{ij} = 0(i, j = 1, \cdots, n)$ and $b_{ij} = 0(i, j = 1, \cdots, n)$, then (3.15) becomes the standard CNN without delay.

In this section, let $(f(x_1(t)), \cdots, f(x_n(t)))$ denote the output of the network at time $t$.

## 5.2 Preliminaries

For any $t_0 \geq 0$, the initial condition of (3.15) is assumed to be

$$\begin{cases} x_i(t) = \phi_i(t), & -\infty < t \leq t_0 \\ \sup\limits_{-\infty < s \leq t_0} |\phi_i(s)| < +\infty, & (i = 1, \cdots, n) \end{cases}$$

where $\phi_i(t)(i = 1, \cdots, n)$ are continuous functions.

A point $X^* = (x_1^*, \cdots, x_n^*)^T \in R^n$ is called an equilibrium of (3.15), if

$$-x_i^* + \sum_{j=1}^{n} (a_{ij} + b_{ij} + c_{ij}) f(x_j^*) + I_i = 0, (i = 1, \cdots, n).$$

Clearly, (3.15) has at least one equilibrium (see Subsection 3.2 for equilibria analysis).

For any $t_0 \geq 0$, denote

$$\|\Phi - X^*\|_{t_0} = \sup_{-\infty < t \leq t_0} \left[ \sum_{i=1}^{n} |\phi_i(t) - x_i^*| \right].$$

DEFINITION 3.4 *Suppose network (3.15) has an equilibrium $X^* = (x_1^*, \cdots, x_n^*)^T$. The equilibrium is called uniformly stable, if for any $t_0 \geq 0$, there is a positive constant $\xi \geq 0$, independent of $t_0$, such that*

$$|x_i(t) - x_i^*| \leq \xi \cdot \|\Phi - X^*\|_{t_0}, \quad (i = 1, \cdots, n)$$

*for all $t \geq t_0$.*

It is clear that this definition of uniform stability is stronger than the usual one, which requires only that for any $t_0 \geq 0$ and any $\epsilon > 0$ there exists a $\delta = \delta(\epsilon) > 0$ such that $\|\Phi - X^*\|_{t_0} < \delta$ implies that

$$|x_i(t) - x_i^*| < \epsilon, \quad (i = 1, \cdots, n)$$

for all $t \geq t_0$.

**DEFINITION 3.5** *An equilibrium $X^* = (x_1^*, \cdots, x_n^*)^T$ for network (3.15) is globally uniformly asymptotically stable, if $X^*$ is uniformly stable, and for any $\epsilon > 0$ and $H > 0$, there exists $T = T(\epsilon, H) > 0$ such that for any $t_0 \geq 0$, $\|\Phi - X^*\|_{t_0} < H$ implies that*

$$|x_i(t) - x_i^*| < \epsilon, \quad i = 1, \cdots, n$$

*for all $t \geq t_0 + T$.*

**DEFINITION 3.6** *An equilibrium $X^* = (x_1^*, \cdots, x_n^*)^T$ for network (3.15) is uniformly output stable, if there is a constant $\xi \geq 0$ such that for any $t_0 \geq 0$,*

$$|f(x_i(t)) - f(x_i^*)| \leq \xi \cdot \|\Phi - X^*\|_{t_0}$$

*for all $t \geq t_0$ and $i = 1, \cdots, n$.*

**DEFINITION 3.7** *An equilibrium $X^* = (x_1^*, \cdots, x_n^*)^T$ for network (3.15) is globally uniformly asymptotically output stable, if it is uniformly output stable, and for any $\epsilon > 0$ and $H > 0$, there exists $T = T(\epsilon, H) > 0$ such that for any $t_0 \geq 0$, $\|\Phi - X^*\|_{t_0} < H$ implies that*

$$|f(x_i(t)) - f(x_i^*)| < \epsilon, \quad (i = 1, \cdots, n)$$

*for all $t \geq t_0 + T$.*

In the above definitions, (3.4) and (3.5) are about the network's state stability, (3.6) and (3.7) are about its output stability. In many applications, it is more common to use output stability than state stability.

## 5.3 Relations Between State Stability and Output Stability

In this subsection, the relationships between the network's state stability and its output stability is analysed.

At any time $t$, the state of (3.15) is $(x_1(t), \cdots, x_n(t))$ and the output of the network is $(f(x_1(t)), \cdots, f(x_n(t)))$. The network's output at a given time is the image of its state at that time under the mapping $f$, i.e.,

$$f : (x_1(t), \cdots, x_n(t)) \to (f(x_1(t)), \cdots, f(x_n(t))).$$

Since $f$ is nonlinear, the relation between the dynamic behaviors of network's state and network's output is not simple. In many practical applications, the dynamic behaviors of network's output are both more interesting than those of the state, and easier to measure. In any case, it is useful to understand the relationships between a network's state stability and its output stability. Such properties are derived below.

THEOREM 3.11 *An equilibrium* $X^* = (x_1^*, \cdots, x_n^*)^T$ *of network (3.15) is uniformly stable if and only if it is uniformly output stable.*

*Proof*: By Lemma 3.3,

$$|f(x_i(t)) - f(x_i^*)| \leq |x_i(t) - x_i^*|, (i = 1, \cdots, n)$$

for all $t \geq 0$. Thus, necessity follows easily. Next, sufficiency will be proven.

Denote

$$\begin{cases} \tilde{x}_i(t) = x_i(t) - x_i^* \\ \tilde{f}(x_i(t)) = f(x_i(t)) - f(x_i^*) \end{cases}$$

for $i = 1, \cdots, n$. Rewrite (3.15) as

$$\begin{aligned} \frac{d\tilde{x}_i(t)}{dt} &= -\tilde{x}_i(t) + \sum_{j=1}^{n} \left[a_{ij}\tilde{f}(x_j(t)) + b_{ij}\tilde{f}(x_j(t - \tau_{ij}(t)))\right] \\ &\quad + \sum_{j=1}^{n} c_{ij} \int_{-\infty}^{t} k_{ij}(s)\tilde{f}(x_j(s))ds, \quad (i = 1, \cdots, n) \quad (3.16) \end{aligned}$$

for $t \geq 0$. Then, from (3.16),

$$\begin{aligned} D^+|\tilde{x}_i(t)| &\leq -|\tilde{x}_i(t)| + \sum_{j=1}^{n} (|a_{ij}| + |b_{ij}|) \cdot \sup_{t-\tau \leq s \leq t} \left(|\tilde{f}(x_i(s))|\right) \\ &\quad + \sum_{j=1}^{n} |c_{ij}| \int_{-\infty}^{t} k_{ij}(s)|\tilde{f}(x_j(s))|ds, \quad (3.17) \end{aligned}$$

for $t \geq 0$ and $(i = 1, \cdots, n)$.

Since $(x_1^*, \cdots, x_n^*)$ is uniformly output stable, for any $t_0 \geq 0$ there is $\xi \geq 1$ such that

$$|\tilde{f}(x_i(t))| \leq \xi \cdot \|\Phi - X^*\|_{t_0}, \quad (i = 1, \cdots, n)$$

for all $t \geq t_0$. From (3.17),

$$D^+|\tilde{x}_i(t)| \leq -|\tilde{x}_i(t)| + \xi \sum_{j=1}^{n} \mu_{ij} \|\Phi - X^*\|_{t_0}$$

for all $t \geq t_0$, where $\mu_{ij} = |a_{ij}| + |b_{ij}| + |c_{ij}|$, and so

$$\begin{aligned} |\tilde{x}_i(t)| &\leq |\tilde{x}_i(t)|e^{-(t-t_0)} \\ &\quad + \xi \sum_{j=1}^{n} \int_{t_0}^{t} e^{-(t-s)} \mu_{ij} \|\Phi - X^*\|_{t_0} ds \\ &\leq \gamma \|\Phi - X^*\|, (i = 1, \cdots, n) \quad (3.18) \end{aligned}$$

for all $t \geq t_0$, where

$$\gamma = 1 + \xi \max_{1 \leq i \leq n} \left( \sum_{j=1}^{n} \mu_{ij} \right) > 1.$$

The equilibrium $X^*$ is thus uniformly stable, and the proof is complete.

THEOREM 3.12 *An equilibrium $X^* = (x_1^*, \cdots, x_n^*)^T$ of network (3.15) is globally uniformly asymptotically stable if and only if it is globally uniformly asymptotically output stable.*

*Proof*: Necessity is easy to prove, as in Theorem 3.11. Since Theorem 3.11 shows that uniform stability of $x^*$ is equivalent to its uniform output stability, to complete the proof of sufficiency, we only need prove that for any $H > 0$, there exists $T = T(\epsilon, H) > 0$ such that for any $t_0 \geq 0$, any $\epsilon > 0$, $\|\Phi - X^*\|_{t_0} < H$ implies that

$$|\tilde{x}_i(t)| < \epsilon, \quad (i = 1, \cdots, n)$$

for all $t \geq t_0 + T$.

It is easy to see from (3.18) that $\|\Phi - X^*\|_{t_0} < H$ implies

$$|\tilde{x}_i(t)| \leq \xi \, \|\Phi - X^*\|_{t_0} \leq \xi H, (i = 1, \cdots, n)$$

for all $t_0 \geq 0$.

Let $\tilde{\mu}_{ij} = |a_{ij}| + |b_{ij}| + 2|c_{ij}|$, and choose a constant $\epsilon_1$ such that

$$0 < \epsilon_1 \leq \min_{1 \leq i \leq n} \left[ \frac{\epsilon}{\xi H + \sum_{j=1}^{n} \tilde{\mu}_{ij}} \right].$$

By the assumption that $X^*$ is globally uniformly asymptotically output stable, there exists $T_1 = T_1(\epsilon_1, H) > 0$ such that for any $t_0 \geq 0$, $\|\Phi - X^*\|_{t_0} < H$ implies

$$|\tilde{f}(x_i(t))| < \epsilon_1, \quad i = 1, \ldots, n$$

for all $t \geq t_0 + T_1$.

Choosing a constant $c \geq \tau$ sufficiently large so that

$$H \cdot \int_{c}^{+\infty} k_{ij}(s) \, ds < \epsilon_1 .$$

Then for any $t \geq t_0 + T_1 + c$, it follows from (3.17) that

$$D^+|\tilde{x}_i(t)| \;\; \leq \;\; -|\tilde{x}_i(t)| + \sum_{j=1}^{n} (|a_{ij}| + |b_{ij}|) \, \epsilon_1$$

$$
\begin{aligned}
& + \sum_{j=1}^{n} |c_{ij}| \left[ \int_{-\infty}^{t-c} k_{ij}(t-s) H ds \right. \\
& \left. + \int_{t-c}^{t} k_{ij}(t-s) \epsilon_1 ds \right] \\
= & -|x_i(t) - x_i^*| + \sum_{j=1}^{n} \left( |a_{ij}| + |b_{ij}| \right) \epsilon_1 \\
& + \sum_{j=1}^{n} |c_{ij}| \left[ H \int_{c}^{+\infty} k_{ij}(s) ds \right. \\
& \left. + \epsilon_1 \int_{0}^{c} k_{ij}(s) ds \right] \\
\leq & -|\tilde{x}_i(t)| + \epsilon_1 \sum_{j=1}^{n} \tilde{\mu}_{ij}
\end{aligned}
$$

and then

$$
\begin{aligned}
|\tilde{x}_i(t)| \leq & \; |\tilde{x}_i(t_0 + T_1)| e^{-(t - T_1 - c - t_0)} \\
& + \epsilon_1 \sum_{j=1}^{n} \int_{t_0 + T_1 + c}^{t} e^{-(t-s)} \tilde{\mu}_{ij} ds \\
\leq & \; \xi H e^{-(t - T_1 - c - t_0)} + \epsilon_1 \sum_{j=1}^{n} \tilde{\mu}_{ij}
\end{aligned}
$$

for all $t \geq t_0 + T_1 + c$.

Choose $T_2 \geq 0$ such that $e^{-T_2} \leq \epsilon_1$, and denote $T = T_1 + T_2 + c$, then

$$
|\tilde{x}_i(t)| \leq \left[ \xi H + \sum_{j=1}^{n} \tilde{\mu}_{ij} \right] \epsilon_1 \leq \epsilon
$$

for all $t \geq t_0 + T$ and $(i = 1, \cdots, n)$. This completes the proof.

The network's state stability is thus equivalent to its output stability.

## 5.4 Global Convergence Analysis

In this section, global convergence conditions for network (3.15) are established. These conditions are quite simple, and are easy to check. They are totally independent of any delays in the network (3.15).

THEOREM 3.13 *Suppose*

$$
-1 + a_{ii} + \sum_{j=1}^{n} \left[ |a_{ij}|(1 - \delta_{ij}) + |b_{ij}| + |c_{ij}| \right] < 0
$$

*for* $i = 1, \cdots, n$, *where*

$$\delta_{ij} = \begin{cases} 1 & \text{if } i = j \\ 0 & \text{if } i \neq j . \end{cases}$$

*Then the network (3.15) has only one equilibrium, which is globally uniformly asymptotically output stable.*

*Proof.* First, an inequality will be derived, which will be used many times in the proof. Clearly, (3.15) has at least one equilibrium $X^* = (x_1^*, \cdots, x_n^*)^T$. Rewrite (3.15) in the form (3.16). From (3.16) by Lemmas 3.3 and 3.4, it follows that

$$\begin{aligned} D^+|\tilde{x}_i(t)| \leq & -|\tilde{x}_i(t)| + a_{ii}|\tilde{f}(x_i(t))| \\ & + \sum_{j=1, j\neq i}^{n} |a_{ij}||\tilde{f}(x_j(t))| \\ & + \sum_{j=1}^{n} |b_{ij}| \sup_{t-\tau\leq s\leq t} \left(|\tilde{f}(x_j(s))|\right) \\ & + \sum_{j=1}^{n} |c_{ij}| \int_{-\infty}^{t} k_{ij}(s)|\tilde{f}(x_j(s))|ds \\ \leq & (-1 + a_{ii})|\tilde{f}(x_i(t))| \\ & + \sum_{j=1}^{n} (|a_{ij}|(1 - \delta_{ij}) + |b_{ij}|) \\ & \times \sup_{t-\tau\leq s\leq t} \left(|\tilde{f}(x_j(s))|\right) \\ & + \sum_{j=1}^{n} |c_{ij}| \int_{-\infty}^{t} k_{ij}(s)|\tilde{f}(x_j(s))|ds, \\ & t \geq 0, (i = 1, \cdots, n). \end{aligned} \tag{3.19}$$

The proof of the theorem is in two parts. In the first part, it will be proved that the equilibrium $X^*$ is uniformly output stable. That is, for any $t_0 \geq 0$

$$|\tilde{f}(x_i(t))| \leq \|\Phi - X^*\|_{t_0}, (i = 1, \cdots, n). \tag{3.20}$$

for all $t \geq t_0$. The proof of this assertion is as follows.

Suppose that $X^*$ is not uniformly output stable. Then there exists some $i$ and $t_1$ and $t_2$ ($t_2 > t_1 \geq t_0$) such that

$$|\tilde{f}(x_i(t_1))| = \|\Phi - X^*\|_{t_0},$$

and $|\tilde{f}(x_i(t))|$ is strictly monotone increasing on the interval $[t_1, t_2]$ (See Figure 3.4 for an intuitive explanation). Moreover,

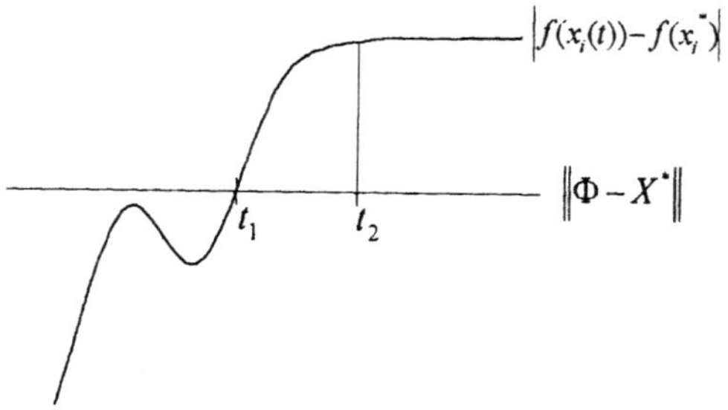


*Figure 3.4.* $|f(x_i(t)) - f(x_i^*)|$ is monotone increasing on interval $[t_1, t_2]$.

$$|\tilde{f}(x_j(s))| \leq \|\Phi - X^*\|_{t_0}, \quad (j = 1, \cdots, n)$$

for all $-\infty < s \leq t_1$. Then, from (3.19), it follows that

$$\begin{aligned} D^+|\tilde{x}_i(t_1)| &\leq \left[-1 + a_{ii} + \sum_{j=1}^{n} \nu_{ij}\right] \|\Phi - X^*\|_{t_0} \\ &< 0 \end{aligned}$$

where $\nu_{ij} = |a_{ij}|(1 - \delta_{ij}) + |b_{ij}| + |c_{ij}|$. By continuity, there must exist a $t_* > t_1 (t_* \leq t_2)$ such that

$$|\tilde{x}_i(t)| \leq |\tilde{x}_i(t_1)|,$$

for all $t \in [t_1, t_*]$. Moreover, by continuity of the solutions of network (3.15), $t_*$ can be chosen so close to $t_1$ such that for any $t \in [t_1, t_*]$, both $x_i(t)$ and $x_i(t_1)$ are in one of the three intervals $[-1, 1]$, $(-\infty, -1]$ and $[1, +\infty)$. Then, by Lemma 3.6,

$$|\tilde{f}(x_i(t))| \leq |\tilde{f}(x_i(t_1))|$$

for all $t \in [t_1, t_*]$. This contradicts the fact that $|\tilde{f}(x_i(t))|$ is strictly monotone increasing on the interval $[t_1, t_2]$, and so (3.20) holds. The proof of uniform output stability is complete.

In the second part of the proof, it will be proved that for any $\epsilon > 0$ and $H > 0$, there is a $T = T(\epsilon, H) > 0$ such that for any $t_0 \geq 0$, $\|\Phi - X^*\|_{t_0} < H$ implies

$$|\tilde{f}(x_i(t))| \leq \epsilon, \quad i = 1, \ldots, n \tag{3.21}$$

for all $t \geq t_0 + T$. This requires several steps.

First, it is clear from (3.20) that $\|\phi - x^*\|_{t_0} < H$ implies $|\tilde{f}(x_i(t))| < H$, for $i = 1, \ldots, n$ and all $t \geq t_0$. By Theorem 3.11, there exists a constant $\xi > 0$ such that $|\tilde{x}_i(t)| < \xi H$, for $i = 1, \ldots, n$ and all $t \geq t_0$.

Define

$$\alpha = \min_{1\leq i\leq n}\left[1 - a_{ii} - \sum_{j=1}^{n}\nu_{ij}\right] > 0$$

and

$$\eta = \frac{\epsilon\cdot\alpha}{2\max_{1\leq i\leq n}\left[\sum_{j=1}^{n}\nu_{ij}\right]}$$

and choose a constant $c \geq \tau$ such that

$$H\int_{c}^{+\infty} k_{ij}(s)\,ds \leq \eta.$$

Let $N$ be the first nonnegative integer such that $\epsilon + N\eta \geq H$, and take

$$t_k = t_0 + kT^*$$

where $k$ is a nonnegative integer and

$$T^* = T^*(\varepsilon, H) = c + \frac{4\xi H}{\alpha\epsilon}.$$

Now, we are in a position to prove (3.21). Firstly, mathematical induction will be used to prove that $\|\Phi - X^*\|_{t_0} < H$ implies

$$|\tilde{f}(x_i(t))| \leq \epsilon + (N-k)\eta, t \geq t_k, (i = 1,\cdots,n) \tag{3.22}$$

for all $k = 0, 1, 2, \cdots, N$. Clearly, this will imply (3.21).

Obviously, (3.22) holds for $k = 0$. Suppose (3.22) holds for some $k(0 \leq k < N)$, *i.e.*,

$$|\tilde{f}(x_i(t))| \leq \epsilon + (N-k)\eta, (i = 1,\cdots,n) \tag{3.23}$$

for all $t \geq t_k$. Next, two steps will be used to prove that

$$|\tilde{f}(x_i(t))| \leq \epsilon + (N-k-1)\eta, (i = 1,\cdots,n) \tag{3.24}$$

for all $t \geq t_{k+1}$.

Firstly, we prove there exists a $\bar{t} \in [t_k + c, t_{k+1}]$ such that

$$|\tilde{f}(x_i(\bar{t}))| \leq \epsilon + (N-k-1)\eta, (i = 1,\cdots,n). \tag{3.25}$$

Otherwise, there exists some $i$ such that

$$|\tilde{f}(x_i(t))| > \epsilon + (N-k-1)\eta \tag{3.26}$$

for all $t \in [t_k + c, t_{k+1}]$. Then, from (3.23) and (3.26),

$$\sup_{t-c\leq\theta\leq t}|\tilde{f}(x_j(\theta))| \leq \epsilon + (N-k)\eta < |\tilde{f}(x_i(t))| + \eta \tag{3.27}$$

for all $t \in [t_k + c, t_{k+1}]$ and $(j = 1, \cdots, n)$. From (3.19), (3.26) and (3.27), it follows for all $t \in [t_k + c, t_{k+1}]$ that

$$\begin{aligned}
D^+|\tilde{x}_i(t)| &\leq (-1 + a_{ii})|\tilde{f}(x_i(t))| \\
&\quad + \sum_{j=1}^{n} (|a_{ij}|(1 - \delta_{ij}) + |b_{ij}|) \\
&\quad \times \left(|\tilde{f}(x_i(t))| + \eta\right) \\
&\quad + \sum_{j=1}^{n} |c_{ij}| \left[\int_{-\infty}^{t-c} k_{ij}(t-s) H \, ds \right. \\
&\quad \left. + \int_{t-c}^{t} k_{ij}(t-s) \left(|\tilde{f}(x_i(t))| + \eta\right) ds \right] \\
&\leq (-1 + a_{ii})|\tilde{f}(x_i(t))| + \sum_{j=1}^{n} |c_{ij}|\eta \\
&\quad + \sum_{j=1}^{n} \nu_{ij} \left(|\tilde{f}(x_i(t))| + \eta\right) \\
&\leq \alpha \left[-|\tilde{f}(x_i(t))| + \frac{\sum_{j=1}^{n} \nu_{ij} + |c_{ij}|}{\alpha}\eta\right] \\
&\leq \alpha(-\epsilon + \frac{\epsilon}{2}) \\
&= -\frac{\alpha}{2}\epsilon
\end{aligned} \tag{3.28}$$

and so

$$\begin{aligned}
|\tilde{x}_i(t_{k+1})| &\leq |\tilde{x}_i(t_k + c)| - \frac{\alpha}{2}\epsilon(T^* - c) \\
&\leq \xi H - 2\xi H \\
&\leq -\xi H.
\end{aligned}$$

This contradiction shows that (3.25) holds.

To complete the proof of (3.24), it is sufficient to prove that

$$|\tilde{f}(x_i(t))| \leq \epsilon + (N - k - 1)\eta, (i = 1, \cdots, n) \tag{3.29}$$

for all $t \geq \bar{t}$. If not, there must exist a $\hat{t} > \bar{t}$ and some $i$ such that

$$|\tilde{f}(x_i(\hat{t}))| > \epsilon + (N - k - 1)\eta$$

and

$$|\tilde{f}(x_i(\hat{t}))| > |\tilde{f}(x_i(t))| \tag{3.30}$$

for all $\bar{t} \leq t \leq \hat{t}$. Then from (3.23)

$$|\tilde{f}(x_j(\theta))| \leq \epsilon + (N-k)\eta < |\tilde{f}(x_i(\hat{t}))| + \eta, (j = 1, \cdots, n)$$

for all $\theta \geq \bar{t}$. Similarly to the calculation of (3.28), it can show that $D^+|\tilde{x}_i(\hat{t})| < 0$. By continuity of the solutions of (3.15), there must exist a $t_{**} \in [\bar{t}, \hat{t}]$ such that both $x_i(t_{**})$ and $x_i(\hat{t})$ are in one of the three intervals $(-\infty, -1]$, $[-1, 1]$ and $[1, +\infty)$, and

$$|\tilde{x}_i(\hat{t})| \leq |\tilde{x}_i(t_{**})|.$$

By Lemma 3.6,

$$|\tilde{f}(x_i(\hat{t}))| \leq |\tilde{f}(x_i(t_{**}))|$$

which contradicts (3.30). This proves (3.29) to be true.

Note that $t_{k+1} \geq \bar{t}$. Then (3.24) follows from (3.29), and so (3.22) holds.

Taking $k = N$ in (3.22), it follows that $||\Phi - X^*||_{t_0} < H$ implies

$$|\tilde{f}(x_i(t))| \leq \epsilon, (i = 1, \cdots, n)$$

for all $t \geq t_0 + T$, where $T = NT^*$ is independent of $t_0$. This proves (3.21).

The above proof shows the equilibrium $X^*$ is globally uniformly asymptotically output stable. In this situation, the network cannot have any other equilibrium. Thus, the proof is completed.

COROLLARY 3.2 *Consider a CNN with variable delays*

$$\begin{aligned} \frac{dx_i(t)}{dt} &= \sum_{j=1}^{n} [a_{ij} f(x_j(t)) + b_{ij} f(x_j(t - \tau_{ij}(t)))] \\ &\quad -x_i(t) + I_i \end{aligned} \tag{3.31}$$

*for* $t \geq 0$ *and* $(i = 1, \cdots, n)$, *where the delays* $\tau_{ij}(t)(i, j = 1, \cdots, n)$ *are assumed to any bounded continuous functions. If*

$$-1 + a_{ii} + \sum_{j=1}^{n} [|a_{ij}|(1 - \delta_{ij}) + |b_{ij}|] < 0, (i = 1, \cdots, n)$$

*then, the network (3.31) has only one equilibrium and this equilibrium is globally uniformly asymptotically stable.*

Obviously, the above stability conditions are independent of the delays. This property is very useful in practical applications since the delays in neural networks are actually not easy to be known exactly in practice. Recently, in [92, 93], stability of neural networks with a variable and differentiable delay was studied. The stability conditions given in [92, 93] depend on the delay. Stability conditions depend on delays are difficult for RNNs to apply in practice.

The conditions given in Theorem 3.13 and Corollary 3.2 contain self inhibitions $a_{ii}(i=1,\cdots,n)$ of neurons. This makes Theorem 3.13 and Corollary 3.2 to be significant. If replace $a_{ii}$ by $|a_{ii}|$ in Theorem 3.13 and Corollary 3.2, the results would be trivial.

## 5.5 Examples

In this section, some examples are given to illustrate the theory.

EXAMPLE 3.4 *Consider a CNN with variable delays*

$$\begin{cases} \dot{x}_1(t) = -x_1(t) - 2f(x_1(t)) + 2f(x_3(t-\tau_1(t))) - \frac{5}{2} \\ \dot{x}_2(t) = -x_2(t) + \frac{1}{2}f(x_2(t-\tau_2(t))) + 1 \\ \dot{x}_3(t) = -x_3(t) + \frac{3}{4}f(x_3(t-\tau_3(t))) + \frac{1}{8}, \quad t \geq 0 \end{cases} \tag{3.32}$$

*for* $t \geq 0$.

In this network, the delays $\tau_i(t) \geq 0(i=1,2,3)$ are assumed to be any bounded continuous functions. Such delays could be uncertain. Using the Corollary 3.2, it is clear that this network has only one equilibrium, which is globally uniformly asymptotically stable. It can be calculated to be at $(-0.5, 1.5, 0.5)^T$. However, the global convergence of (3.32) cannot be checked by any previously known criteria, say, stability conditions in [92, 93]. Simulations of the convergence of (3.32) in the case of $\tau_i(t) = 0(i=1,2,3)$ are shown in Figure 3.5.

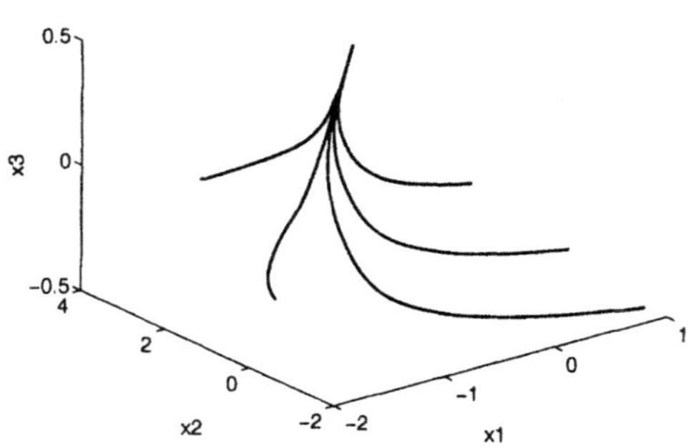


*Figure 3.5.* $3D$ phase space $(x_1, x_2, x_3)$

EXAMPLE 3.5 *Consider a two dimensional CNN with infinite delay*

$$\dot{x}_1(t) \quad = \quad -x_1(t) + \frac{1}{3}f\left(x_1\left(t - \sin^2(t)\right)\right)$$

$$
\begin{aligned}
& + \int_{-\infty}^{t} \frac{1}{\pi[1+(t-s)^2]} \cdot f(x_2(s))ds - \frac{13}{6} \\
\dot{x}_2(t) &= -x_2(t) + \frac{1}{4} f\left(x_2\left(t - \cos^2(t)\right)\right) \\
& + \int_{-\infty}^{t} e^{-2(t-s)} \cdot f\left(x_1(s)\right) ds + \frac{9}{4}
\end{aligned}
$$

*for* $t \geq 0$.

Noting that

$$
\begin{aligned}
\int_0^{+\infty} \frac{1}{1+s^2} ds &= \frac{\pi}{2} \\
\int_0^{+\infty} e^{-2s} ds &= \frac{1}{2},
\end{aligned}
$$

the stability conditions in Theorem 3.13 are satisfied. Then this network has one unique equilibrium and this equilibrium is globally uniformly asymptotically output stable. The equilibrium can be found out to be $(-2, 2)$.

## 6. Conclusions

In this chapter, convergence of CNNs have been studied. Complete stability of standard CNNs is re-analyzed. By relating the standard CNNs with another differential equation which possess some good properties, the non-differentiable problem encountered in complete stability analysis of output function of CNNs has been overcomed. For more results on complete stability of symmetric CNNs, refer to [59, 124, 39, 41, 67], for non-symmetric CNNs, refer to [39, 180, 181, 179, 13].

Global exponential stability and global exponential convergent rate estimation are discussed for standard CNNs via constructing Lyapunov functions and inequalities analysis. Simple conditions for GES are established and some convergent rate are obtained. For more results on standard CNNs, refer to [199, 95, 121, 226]

By constructing an interesting Lyapunov functional which contains some special terms, conditions of GES for CNNs with constant delays are established. Results regarding global convergence of CNNs with constant delays, refer to [9, 119, 120].

It is well known that delays often appear in artificial neural networks, though to identify them is not easy: apparently constant delay is only an ideal and simplified case. In most situations, delays are variable and may extend over all the past. A model of CNNs which contains variable and infinite delays is proposed. Convergence is analyzed. This model may have applications in processing of motion related phenomena such as moving images and associative memories.

CNNs have been extensively studied in recent years. A lot of results are reported. For more dynamical results obtained on CNNs, please refer to [129, 146, 80, 79, 5, 6, 24, 167, 144, 186, 185, 47, 164]. For more applications of CNNs, refer to [7, 175, 18].

Chapter 4

# RECURRENT NEURAL NETWORKS WITH UNSATURATING PIECEWISE LINEAR ACTIVATION FUNCTIONS

## 1. Introduction

It is known that in the dynamical analysis of RNNs, the activation functions are important factors which affect the dynamics of neural networks. Various activation functions have been used for neural networks. Recently, the studies reported in [19, 51, 75, 76, 203, 204, 208, 54] are focused on a class of RNNs with unsaturating linear threshold activation functions (LT networks). This class of neural networks has potential in many important applications. In [76], an efficient silicon design for LT networks has been demonstrated, with a discussion on the co-existence of analog amplification and digital selection in networks circuits. This transfer function is also more appropriate for RNNs [51]. Since the unsaturating piecewise linear activation function are unbounded, more complex dynamic properties may exist in the networks.

The model of RNNs with unsaturating piecewise linear activation functions is described by the following nonlinear differential equation:

$$\dot{x}_i(t) = -x_i(t) + \sum_{j=1}^{n} w_{ij}\sigma(x_j(t)) + h_i, \quad (i = 1, \cdots, n), \tag{4.1}$$

or its equivalent vector form

$$\dot{x}(t) = -x(t) + W\sigma(x(t)) + h$$

for all $t \geq 0$, where each $x_i$ denotes the activity of neuron $i$, $W = (w_{ij})_{n \times n}$ is a real $n \times n$ matrix, each of its elements $w_{ij}$ denotes a synaptic weight and represents the strength of the synaptic connection from neuron $j$ to neuron $i$, $x = (x_1, \cdots, x_n)^T \in R^n$, $h \in R^n$ denotes external inputs. For any $x \in R^n$, $\sigma(x) = (\sigma(x_1), \cdots, \sigma(x_n))^T$, and the activation function $\sigma$ is defined as follows:

$$\sigma(s) = \max\{0, s\}, \quad s \in R.$$

The activation function $\sigma$ is an unsaturating piecewise linear function, which is continuous, unbounded and non-differentiable. Figure 4.1 shows this function.

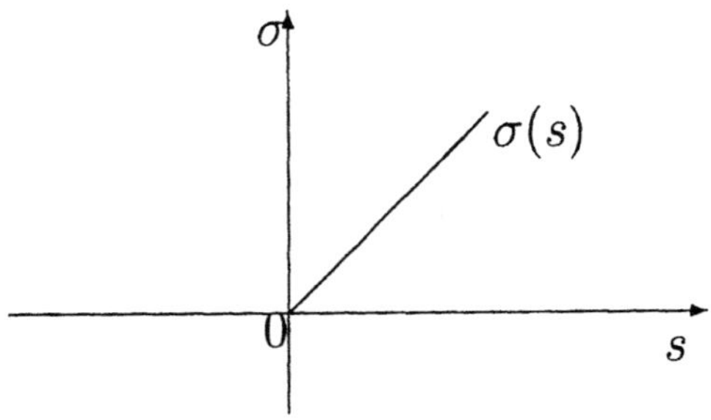


*Figure 4.1.* The activation function $\sigma(s)$

The dynamical properties of RNNs play primarily important roles in their applications. The dynamical analysis of RNNs, especially convergence analysis, has attracted extensive interests and attention of scientists in recent years. Many of the reported conditions for the convergence of neural networks are centered around monostability. However, monostable networks are computationally restrictive: they cannot deal with important neural computations [75], such as those necessary in decision making, where multistable networks become necessary. Recently, there has been increasing interest in multistability analysis for neural networks (see for examples, [75, 76, 203].

Multistable networks are significantly different from monostable ones, and their analysis also merit different treatment. Monostable networks restrict each network to one equilibrium point, whereas a multistable network can have multiple equilibrium points. Conditions for monostability are usually not applicable to multistable networks. In our view, there are three basic properties of a multistable network. The first property is a boundedness characteristic. It is interesting to derive conditions that not only guarantee boundedness, but also allow networks to possess more than one equilibrium point. The second property is that of attractivity. It is necessary to establish whether there exist compact sets into which all the trajectories of a network will be attracted and bounded indefinitely. Clearly, all the equilibrium points and limit circles (if any) must be located in the attracting compact sets. The third property relates to complete convergence. It is necessary to know if every trajectory of a network converges to the equilibrium set.

The above mentioned three dynamical properties for the network (4.1) will be addressed. Conditions based on local inhibition will be derived to guarantee boundedness, and also allow the network to have multi-equilibrium points. Under such boundedness conditions, it will show that the network possesses compact sets which attract all the trajectories of the network. Explicit inequalities to calculate the global attractive sets will be given. In addition, if the weight matrix of the network possesses some kind of symmetry, it will show the network (4.1) has the property of complete convergence.

Recently in [203], it studied the boundedness of the following neural networks:

$$\dot{y}(t) = -y(t) + \sigma\left(Wy(t) + h\right) \tag{4.2}$$

for $t \geq 0$. Based on local inhibition of the network, [203] established some interesting and elegant boundedness conditions. The network (4.2) has some difference from the network described in (4.1). Dynamically, the network (4.1) is more general than that of (4.2). In fact, if we let $x(t) = Wy(t) + h$, then (4.2) can be easily transformed to (4.1). The networks (4.1) and (4.2) have the following relationship:

$$\dot{y}(t) = -y(t) + \sigma(x(t))$$

for $t \geq 0$. For any $T \geq 0$, it follows that

$$\|y(t)\| \leq \|y(T)\|e^{-(t-T)} + \int_T^t e^{-(t-s)}\|\sigma(x(s))\|ds, \tag{4.3}$$

for $t \geq T$. By (4.3), the boundedness, attractivity and convergence of (4.1) can be easily translated to network (4.2). However, network (4.1) cannot be transformed into (4.2) and retain the dynamic properties, except when matrix $W$ is invertible [203]. In fact, if $W$ is not invertible, the equilibrium points of (4.1) and (4.2) do not have a one-to-one correspondence under the above transform. In many applications it may not be reasonable to assume that the matrix $W$ is invertible. Many neural systems exhibiting short-term memory are modeled by non-invertible networks, such as the oculomotor integrator [163] or the head-direction system [230]. In these networks, precise tuning of synaptic connections is necessary to result in line attractors along which the dynamics are marginally stable. In addition, the attractivity issue has not been studied in [203]. It is believed that attractivity is an important property for multistable networks.

## 2. Preliminaries

Since the activation function $\sigma(\cdot)$ is an important factor which affects the dynamics of the network, important properties which will be useful in the subsequent multistability analysis will first be presented.

LEMMA 4.1

$$s\sigma(s) = \sigma^2(s) = 2\int_0^s \sigma(\theta)d\theta,$$

*for any* $s \in R$.

*Proof*: If $s \geq 0$, then

$$2\int_0^s \sigma(\theta)d\theta = 2\int_0^s \theta d\theta = s^2 = \sigma^2(s) = s\sigma(s).$$

If $s < 0$, then $\sigma(\theta) = 0$ for all $s \leq \theta \leq 0$. Thus,

$$2\int_0^s \sigma(\theta)d\theta = 0 = s\sigma(s) = \sigma^2(s).$$

This completes the proof.

LEMMA 4.2

$$[\sigma(s) - \sigma(\theta)]^2 \leq (s-\theta)[\sigma(s) - \sigma(\theta)]$$

*for all* $s, \theta \in R$.

*Proof*: For any $s, \theta \in R$, clearly,

$$\sigma(s) - \sigma(\theta) = \sigma(s - \theta + \theta) - \sigma(\theta) \leq \sigma(s-\theta).$$

Without loss of generality, assume that $s \geq \theta$. Then $\sigma(s) \geq \sigma(\theta)$, and so

$$[\sigma(s) - \sigma(\theta)]^2 \leq \sigma(s-\theta)[\sigma(s) - \sigma(\theta)] \leq (s-\theta)[\sigma(s) - \sigma(\theta)].$$

This completes the proof.

Denote in this chapter a norm in $R^n$ by

$$\|x\| = \sqrt{\sum_{i=1}^n |x_i|^2},$$

for any $x \in R^n$.

Then, from Lemma 4.1 and Lemma 4.2, it follows that

$$x^T\sigma(x) = \sigma^T(x)\sigma(x) = 2\sum_{i=1}^n \int_0^{x_i} \sigma(\theta)d\theta,$$

and

$$\|\sigma(x) - \sigma(y)\|^2 \leq (x-y)^T \cdot [\sigma(x) - \sigma(y)],$$

for all $x, y \in R^n$

DEFINITION 4.1 *The network (4.1) is bounded if each of its trajectories is bounded.*

DEFINITION 4.2 *Let $S$ be a compact subset of $R^n$. Denote the $\epsilon$-neighborhood of $S$ by $S_\epsilon$. The compact set $S$ is called a global attractive set of network (4.1), if for any $\epsilon > 0$, all trajectories of (4.1) ultimately enter and remain in $S_\epsilon$.*

Let $f(t)$ be a continuous function which is defined on $[0, +\infty)$, the upper limit of $f(t)$ is defined as

$$\lim_{t \to +\infty} \sup f(t) = \lim_{t \to +\infty} \left[ \sup_{\theta \geq t} f(\theta) \right].$$

Clearly, if $f(t)$ is upper bounded, its upper limit exists. If $\lim_{t \to +\infty} \sup f(t) = \xi < +\infty$, then for any $\epsilon > 0$, there exists a $T \geq 0$ such that

$$f(t) \leq \xi + \epsilon$$

for $t \geq T$.

If every trajectory $x(t)$ of (4.1) satisfies

$$\lim_{t \to +\infty} \sup |x_i(t)| \leq s_i < +\infty,$$

then, clearly,

$$S = \{x \| x_i| \leq s_i, (i = 1, \cdots, n)\}$$

is a global attractive set of (4.1).

A vector $x^* \in R^n$ is called an equilibrium point of (4.1), if it satisfies

$$-x^* + W\sigma(x^*) + h \equiv 0.$$

We use $S^e$ to denote the equilibrium set of (4.1).

DEFINITION 4.3 *The network (4.1) is said to be completely stable, if each trajectory converges to an equilibrium point.*

Throughout this paper, for any constant $c \in R$, denote $c^+ = \max(0, c)$. Also, denote

$$h^+ = \left(h_1^+, \cdots, h_n^+\right)^T.$$

For any matrix $A = (a_{ij})_{n \times n}$, denote $A^+ = \left(a_{ij}^+\right)$, where

$$a_{ij}^+ = \begin{cases} a_{ii}, & i = j, \\ \max(0, a_{ij}), & i \neq j. \end{cases}$$

Denote the smallest and largest eigenvalues of $A$ by $\lambda_{min}(A)$ and $\lambda_{max}(A)$, respectively.

# 3. Multistability Analysis

## 3.1 Boundedness and Global Attractivity

In this section, conditions based on local inhibition will be derived which guarantee boundedness and allow the network (4.1) to have property of multistability. Moreover, explicit expressions will be provided, locating the compact sets which globally attract all the trajectories of the network (4.1).

Firstly, a useful lemma is established.

LEMMA 4.3 *Let $M \geq 0$ and $T \geq 0$ be constants. If $x_i(t) \leq M, (i = 1, \cdots, n)$ for $t \geq T$, then we have*

$$|x_i(t)| \leq |x_i(T)|e^{-(t-T)} + M\sum_{j=1}^{n}|w_{ij}| + |h_i|, (i = 1, \cdots, n),$$

*for all $t \geq T$.*

*Proof*: Since $x_i(t) \leq M(i = 1, \cdots, n)$ for $t \geq T$, then by the definition of the activation function $\sigma(\cdot)$, clearly it holds that $\sigma(x_i(t)) \leq M$ for all $t \geq T$. From (4.1),

$$x_i(t) = x_i(T)e^{-(t-T)} + \int_T^t e^{-(t-s)}\left[\sum_{j=1}^{n} w_{ij}\sigma(x_j(s)) + h_i\right] ds, (i = 1, \cdots, n),$$

for $t \geq T$. Then, it follows that

$$|x_i(t)| \leq |x_i(T)|e^{-(t-T)} + M\sum_{j=1}^{n}|w_{ij}| + |h_i|, (i = 1, \cdots, n)$$

for all $t \geq T$. This completes the proof.

The above Lemma shows that if a trajectory of (4.1) is upper bounded then it is lower bounded also. This property will help us to prove the following theorem.

THEOREM 4.1 *If there exist constants $\alpha_i > 0(i = 1, \cdots, n)$ such that*

$$\gamma_i \triangleq 1 - w_{ii} - \frac{1}{\alpha_i}\sum_{j=1, j\neq i}^{n} \alpha_j w_{ij}^{+} > 0, \quad (i = 1, \cdots, n),$$

*then, the network (4.1) is bounded. Moreover,*

$$S = \left\{ x \,\middle|\, |x_i| \leq \Pi \cdot \sum_{j=1}^{n}|w_{ij}| + |h_i|, (i = 1, \cdots, n) \right\},$$

*is a global attractive set of the network (4.1), where*

$$\Pi = \max_{1 \leq j \leq n} \{\alpha_j\} \cdot \max_{1 \leq j \leq n} \left\{ \frac{h_j^+}{\gamma_j} \right\}.$$

*Proof*: Since $\sigma(\cdot) \geq 0$, it follows from (4.1) that

$$\dot{x}_i(t) \leq -x_i(t) + w_{ii}\sigma(x_i(t)) + \sum_{j=1, j \neq i}^{n} w_{ij}^+ \sigma(x_j(t)) + h_i^+, (i = 1, \cdots, n), \tag{4.4}$$

for $t \geq 0$.

Define

$$v_i(t) = \frac{x_i(t)}{\alpha_i}, (i = 1, \cdots, n),$$

for $t \geq 0$. Then, from (4.4),

$$\dot{v}_i(t) \leq -v_i(t) + w_{ii}\sigma(v_i(t)) + \frac{1}{\alpha_i} \left( \sum_{j=1, j \neq i}^{n} \alpha_j w_{ij}^+ \sigma(v_j(t)) + h_i^+ \right), (i = 1, \cdots, n), \tag{4.5}$$

for $t \geq 0$.

For any $x(0) \in R^n$, denote

$$m = \|x(0)\| \cdot \max_{1 \leq i \leq n} \left\{ \frac{1}{\alpha_i} \right\} + \max_{1 \leq i \leq n} \left\{ \frac{h_i^+}{\alpha_i \gamma_i} \right\} + 1.$$

It will prove that

$$v_i(t) < m, (i = 1, \cdots, n), \tag{4.6}$$

for all $t \geq 0$. Otherwise, since $v_i(0) < m (i = 1, \cdots, n)$, then there must exist some $i$ and a time $t_1 > 0$ such that $v_i(t_1) = m$ and

$$v_j(t) \begin{cases} < m, & j = i, \quad 0 \leq t < t_1 \\ \leq m, & j \neq i, \quad 0 \leq t \leq t_1. \end{cases}$$

Thus, it must follow that $\dot{v}_i(t_1) \geq 0$. However, from (4.5), it follows that

$$\begin{aligned} \dot{v}_i(t_1) &\leq -m + w_{ii}m + \frac{1}{\alpha_i} \left( \sum_{j=1, j \neq i}^{n} \alpha_j w_{ij}^+ m + h_i^+ \right) \\ &= -m \cdot \gamma_i + \frac{h_i^+}{\alpha_i} \\ &< 0. \end{aligned}$$

This is a contradiction which proves (4.6) to be true. From (4.6), it follows that

$$x_i(t) < \max_{1\leq j\leq n} \{\alpha_j\} \cdot m \triangleq M, (i = 1, \cdots, n),$$

for all $t \geq 0$. By Lemma 4.3,

$$\begin{aligned} |x_i(t)| &\leq |x_i(0)|e^{-t} + M \cdot \sum_{j=1}^{n} |w_{ij}| + |h_i| \\ &\leq |x_i(0)| + M \cdot \sum_{j=1}^{n} |w_{ij}| + |h_i|, (i = 1, \cdots, n), \end{aligned}$$

for $t \geq 0$. This proves that the network (4.1) is bounded.

Next, it will prove the global attractivity under the conditions. Denote

$$\lim_{t\to+\infty} \sup v_i(t) = \xi_i, (i = 1, \cdots, n).$$

Since each $v_i(t)$ is upper bounded, then $\xi_i < +\infty, (i = 1, \cdots, n)$. It will prove that

$$\xi_i < \max_{1\leq j\leq n} \left\{ \frac{h_j^+}{\gamma_j} \right\}, (i = 1, \cdots, n). \tag{4.7}$$

For the sake of convenience in the subsequent discussion, and without loss of generality, assume that $\xi_l = \max_{1\leq j\leq n} \{\xi_j\}$.

If $\xi_l \leq 0$, clearly (4.7) is true. In what follows, consider the case that $\xi_l > 0$. Suppose (4.7) is not true, i.e.,

$$\xi_l > \max_{1\leq j\leq n} \left\{ \frac{h_j^+}{\gamma_j} \right\}. \tag{4.8}$$

It will prove that this assumption will lead to a contradiction which will then confirm that (4.7) is true.

Clearly, it can choose a small $\epsilon > 0$ such that

$$\xi_l > \max_{1\leq j\leq n} \left\{ \epsilon, \frac{2\epsilon\left(1 + |w_{jj}|\right) + h_j^+}{\gamma_j} \right\} \triangleq k.$$

For this choice of $\epsilon$, by the basic property of the upper limit, there exists a $t_2 \geq 0$ such that

$$v_j(t) \leq \xi_j + \epsilon \leq \xi_l + \epsilon, (j = 1, \cdots, n),$$

for all $t \geq t_2$.

It will prove that there exists a $t_3 \geq t_2$ such that

$$\dot{v}_l(t) < 0 \tag{4.9}$$

for all $t \geq t_3$. If (4.9) is not true, there must exist a $t_4 > t_2$ such that

$$\dot{v}_l(t_4) \geq 0, \quad v_l(t_4) \geq \xi_l - \epsilon.$$

However, from (4.5), it follows that

$$\begin{aligned}
\dot{v}_l(t_4) &\leq -(\xi_l - \epsilon) + w_{ll}\xi_l + |w_{ll}|\epsilon \\
&\quad + \frac{1}{\alpha_l}\sum_{j=1, j\neq i}^{n} \alpha_j w_{lj}^{+}\sigma(\xi_j + \epsilon) + h_l^{+} \\
&\leq -(\xi_l - \epsilon) + w_{ll}\xi_l + |w_{ll}|\epsilon \\
&\quad + \frac{1}{\alpha_l}\sum_{j=1, j\neq i}^{n} \alpha_j w_{lj}^{+}(\xi_l + \epsilon) + h_l^{+} \\
&\leq -\gamma_l\xi_l + 2\epsilon\,(1 + |w_{ll}|) + h_l^{+} \\
&\leq -\gamma_l k + 2\epsilon\,(1 + |w_{ll}|) + h_l^{+} \\
&< 0,
\end{aligned}$$

which is a contradiction, thus proving that (4.9) is true. By (4.9), it shows that $v_l(t)$ is monotonically decreasing. Thus, the limit of $v_l(t)$ exists, that is

$$\lim_{t\to+\infty} v_l(t) = \lim_{t\to+\infty} \sup v_l(t) = \xi_l.$$

Then, there exists a $t_5 \geq t_3$ such that

$$\begin{cases} \xi_l - \epsilon < v_l(t) < \xi_l + \epsilon \\ v_j(t) < \xi_j + \epsilon < \xi_l + \epsilon, (j = 1, \cdots, n) \end{cases}$$

for all $t \geq t_5$.

From (4.5), it follows for $t \geq t_5$ that

$$\begin{aligned}
\dot{v}_l(t) &\leq -(\xi_l - \epsilon) + w_{ll}\xi_l + |w_{ll}|\epsilon \\
&\quad + \frac{1}{\alpha_l}\sum_{j=1, j\neq i}^{n} \alpha_j w_{lj}^{+}\sigma(\xi_l + \epsilon) + h_l^{+} \\
&\leq -\gamma_l\xi_l + 2\epsilon\,(1 + |w_{ll}|) + h_l^{+} \\
&\leq -\gamma_l k + 2\epsilon\,(1 + |w_{ll}|) + h_l^{+}
\end{aligned}$$

Thus

$$\begin{aligned}
v_l(t) &\leq v_l(t_5) - (t - t_5)\left[\gamma_l k - 2\epsilon\,(1 + |w_{ll}|) - h_l^{+}\right] \\
&\to -\infty, \quad \text{as } t \to +\infty.
\end{aligned}$$

This contradicts to $\xi_l > 0$. Therefore, (4.8) does not hold and instead, (4.7) is true.

By (4.7),

$$\lim_{t\to+\infty} \sup x_i(t) = \alpha_i \xi_i \leq \max_{1\leq j\leq n} \{\alpha_j\} \cdot \max_{1\leq j\leq n} \left\{ \frac{h_j^+}{\gamma_j} \right\} \triangleq \Pi, (i = 1, \cdots, n).$$

For any $\eta > 0$, there exists a $t_6 \geq 0$ such that

$$x_i(t) \leq \Pi + \eta, \quad (i = 1, \cdots, n),$$

for $t \geq t_6$. By Lemma 4.3, it follows that

$$|x_i(t)| \leq |x_i(t_6)| e^{-(t-t_6)} + (\Pi + \eta) \cdot \sum_{j=1}^{n} |w_{ij}| + |h_i|,$$

for $t \geq t_6$. Thus,

$$\lim_{t\to+\infty} \sup |x_i(t)| \leq (\Pi + \eta) \cdot \sum_{j=1}^{n} |w_{ij}| + |h_i|.$$

Letting $\eta \longrightarrow 0$, it follows that

$$\lim_{t\to+\infty} \sup |x_i(t)| \leq \Pi \cdot \sum_{j=1}^{n} |w_{ij}| + |h_i|, (i = 1, \cdots, n).$$

This shows that $S$ is a global attractive set of (4.1). The proof is completed.

The local inhibition based conditions of Theorem 4.1 are identical to [203] for neural network (4.2).

THEOREM 4.2 *If*

$$\Omega = I - \left[ \frac{W^T + W}{2} \right]^+$$

*is a positive definite matrix, then the network (1) is bounded. Moreover,*

$$S = \left\{ x \,\middle|\, \|x\| \leq \sqrt{\lambda_{max}\left((W^T W)^+\right)} \cdot \frac{\|h^+\|}{\lambda_{min}(\Omega)} + \|h\| \right\}$$

*is a global attractive set of the network (4.1).*

*Proof*: Define a differentiable function

$$V(t) = \sum_{j=1}^{n} \int_0^{x_i(t)} \sigma(\theta) d\theta,$$

for $t \geq 0$. Obviously, $V(t) \geq 0$ for all $t \geq 0$.

Calculating the derivative of $V(t)$ along the trajectories of (4.1), it follows that

$$\begin{aligned}
\dot{V}(t) &= -\sigma^T(x(t))x^T(t) + \sigma^T(x(t))W\sigma(x(t)) + \sigma^T(x(t))h \\
&\leq \sigma^T(x(t))\left[\frac{W^T+W}{2} - I\right]\sigma(x(t)) + \sigma^T(x(t))h^+ \\
&\leq \sigma^T(x(t))\left(\left[\frac{W^T+W}{2}\right]^+ - I\right)\sigma(x(t)) + \sigma^T(x(t))h^+ \\
&\leq -\lambda_{min}(\Omega)\cdot\sigma^T(x(t))\sigma(x(t)) + \sigma^T(x(t))h^+ \\
&\leq -\frac{\lambda_{min}(\Omega)}{2}\sigma^T(x(t))\sigma(x(t)) + \frac{\|h^+\|^2}{2\lambda_{min}(\Omega)}
\end{aligned}$$

for all $t \geq 0$. By Lemma 4.1,

$$\dot{V}(t) \leq -\lambda_{min}(\Omega)\,V(t) + \frac{\|h^+\|^2}{2\lambda_{min}(\Omega)},$$

for all $t \geq 0$, and

$$\begin{aligned}
V(t) &\leq V(0)e^{-\lambda_{min}(\Omega)t} + \frac{\|h^+\|^2}{2\lambda_{min}(\Omega)}\int_0^t e^{-\lambda_{min}(\Omega)(t-s)}ds \\
&\leq V(0)e^{-\lambda_{min}(\Omega)t} + \frac{\|h^+\|^2}{2\left[\lambda_{min}(\Omega)\right]^2},
\end{aligned}$$

for all $t \geq 0$. By Lemma 4.1 again,

$$V(0) = \sum_{j=1}^{n}\int_0^{x_i(0)}\sigma(\theta)d\theta = \frac{x^T(0)\sigma(x(0))}{2} \leq \frac{\|x(0)\|^2}{2},$$

and

$$\|\sigma(x(t))\|^2 \leq \|x(0)\|^2 e^{-\lambda_{min}(\Omega)t} + \frac{\|h^+\|^2}{\left[\lambda_{min}(\Omega)\right]^2} \triangleq \phi(t), \tag{4.10}$$

for all $t \geq 0$. Moreover,

$$\lim_{t\to+\infty}\sqrt{\phi(t)} = \frac{\|h^+\|}{\lambda_{min}(\Omega)}. \tag{4.11}$$

From (4.1),

$$x(t) = x(0)e^{-t} + \int_0^t e^{-(t-s)}\left[W\sigma(x(s)) + h\right]ds, \tag{4.12}$$

for $t \geq 0$. Then, from (4.12) and (4.10),

$$\begin{aligned}
\|x(t)\| &\leq \|x(0)\|e^{-t} + \int_0^t e^{-(t-s)}\|W\sigma(x(s))\|ds + \|h\| \\
&\leq \|x(0)\|e^{-t} + \int_0^t e^{-(t-s)}\sqrt{\sigma^T(x(s))\left(W^TW\right)\sigma(x(s))}ds + \|h\| \\
&\leq \|x(0)\|e^{-t} + +\|h\| \\
&\quad + \int_0^t e^{-(t-s)}\sqrt{\sigma^T(x(s))\left((W^TW)^+\right)\sigma(x(s))}ds \\
&\leq \|x(0)\|e^{-t} + +\|h\| \\
&\quad + \sqrt{\lambda_{max}\left((W^TW)^+\right)} \cdot \int_0^t e^{-(t-s)}\|\sigma(x(s))\|ds \\
&\leq \|x(0)\|e^{-t} + \sqrt{\lambda_{max}\left((W^TW)^+\right)} \cdot \int_0^t e^{-(t-s)}\sqrt{\phi(s)}ds + \|h\| \\
&\triangleq \Phi(t),
\end{aligned}$$

for $t \geq 0$.

Since

$$\begin{aligned}
\lim_{t\to+\infty}\Phi(t) &= \sqrt{\lambda_{max}\left((W^TW)^+\right)}\lim_{t\to+\infty}\int_0^t e^{-(t-s)}\sqrt{\phi(s)}ds + \|h\| \\
&= \sqrt{\lambda_{max}\left((W^TW)^+\right)}\lim_{t\to+\infty}\sqrt{\phi(t)} + \|h\|,
\end{aligned}$$

then, using (4.11), it follows that

$$\lim_{t\to+\infty}\Phi(t) = \sqrt{\lambda_{max}\left((W^TW)^+\right)} \cdot \frac{\|h^+\|}{\lambda_{min}(\Omega)} + \|h\|.$$

Thus,

$$\lim_{t\to+\infty}\sup\|x(t)\| \leq \sqrt{\lambda_{max}\left((W^TW)^+\right)} \cdot \frac{\|h^+\|}{\lambda_{min}(\Omega)} + \|h\|.$$

This shows that $S$ globally attracts the network (4.1). The proof is completed.

In Theorem 4.1 and Theorem 4.2, if $h_i \leq 0$ $(i = 1, \cdots, n)$, the global attractive sets become

$$S = \{x \,|\, |x_i| \leq |h_i|, (i = 1, \cdots, n)\},$$

and

$$S = \{x \,|\, \|x\| \leq \|h\|\},$$

in Theorem 4.1 and 4.2, respectively. In particular, if $h_i = 0$ $(i = 1, \cdots, n)$, the attracting sets contain only the origin, i.e., the network (4.1) can have only the origin as its unique equilibrium point and all trajectories of (4.1) converge to the origin.

The conditions attached to the weight matrix of the network (4.1) in Theorem 4.1 and Theorem 4.2 are different. For example, consider the matrices

$$W_1 = \begin{bmatrix} 0 & 6 & 7 \\ -4 & 0 & 5 \\ -5 & -3 & 0 \end{bmatrix}, \quad W_2 = \begin{bmatrix} 0 & 5 & 0.5 \\ -7 & 0 & 2 \\ 0.5 & -4 & 0 \end{bmatrix}.$$

Matrix $W_1$ satisfies the conditions in Theorem 4.1, but it does not satisfy the conditions given in Theorem 4.2. On the other hand, matrix $W_2$ satisfies conditions in Theorem 4.2, but not the conditions of Theorem 4.1. Hence, while both theorems relate to boundedness and global attractivity, they can be used complementarily in different situations.

## 3.2 Complete Stability

In this section, the complete stability of (4.1) will be studied, assuming that the connection matrix $W$ is symmetric in some sense.

The general method to study complete convergence of neural networks is to construct suitable energy functions. However, for the network (4.1), since the activation function $\sigma(\cdot)$ is not differentiable, it is difficult to construct energy functions such as those used in [42, 45, 85, 203, 205]. To counter this non-differentiability problem, associate the network (4.1) with another differential equation. Through this equation, a novel energy-like function can be constructed and the conditions for complete stability derived accordingly.

For any $x_0 \in R^n$, associate the network (4.1) with the following equivalent differential equation instead:

$$\dot{z}(t) = -z(t) + \sigma\left(Wz(t) + h + (x_0 - h)e^{-t}\right) \qquad (4.13)$$

for $t \geq 0$.

LEMMA 4.4 *For any $x_0 \in R^n$, the solution $x(t)$ of (4.1) starting from $x_0$ can be represented as*

$$x(t) = Wz(t) + h + (x_0 - h)e^{-t},$$

*where $z(t)$ is the solution of (4.13) starting from the origin.*

*Proof*: For any $x_0 \in R^n$, let $x(t)$ be the solution of (4.1) starting from $x_0$. Consider the following differential equation:

$$\dot{z}(t) = -z(t) + \sigma(x(t)), \qquad (4.14)$$

for all $t \geq 0$.

From (4.1) and (4.14),

$$\frac{d\left[x(t)-h-Wz(t)\right]}{dt}=-\left[x(t)-h-Wz(t)\right],$$

for $t \geq 0$. Solving this equation, it follows that

$$x(t)=(x_0-h-Wz(0))e^{-t}+Wz(t)+h,$$

for $t \geq 0$, where $z(0) \in R^n$ is any initial vector of (4.14). Taking $z(0)=0$ in particular, i.e., $z(t)$ is the solution of (4.14) starting from the origin, then

$$x(t)=(x_0-h)e^{-t}+Wz(t)+h, \tag{4.15}$$

for $t \geq 0$. The result now follows by substituting (4.15) into (4.14). The proof is completed.

LEMMA 4.5 *For any $x_0 \in R^n$, the solution $z(t)$ of (4.13) starting from the origin satisfies*

$$z_i(t) \geq 0, (i=1,\cdots,n)$$

*for all $t \geq 0$.*

*Proof*: Since $z(0)=0$, then from (4.13),

$$z_i(t)=\int_0^t e^{-(t-s)}\sigma\left(\sum_{j=1}^n w_{ij}z_j(s)+h_i+(x_0^i-h_i)e^{-s}\right)ds,$$

for $t \geq 0$ and $(i=1,\cdots,n)$. Since $\sigma(\cdot) \geq 0$ always, clearly, $z_i(t) \geq 0$ for $t \geq 0$. This completes the proof.

LEMMA 4.6 *Suppose there exists a diagonal positive definite matrix $D$ such that the matrix $DW$ is symmetric. Given any $x_0 \in R^n$, if the solution $z(t)$ of (4.13) starting from the origin is bounded, then*

$$\lim_{t\to+\infty}\dot{z}(t)=0.$$

*Proof*: Constructing an energy-like function

$$\begin{aligned} E(t) &= \frac{1}{2}z^T(t)D(I-W)z(t)-h^TDz(t) \\ &\quad -(x_0-h)^TD\left(z(t)e^{-t}+\int_0^t z(s)e^{-s}ds\right), \end{aligned} \tag{4.16}$$

for all $t \geq 0$. Clearly, $E(0)=0$.

By the assumption that $z(t)$ is bounded, it is easy to see that $E(t)$ is also bounded. By Lemma 4.5, $z_i(t) \geq 0 (i=1,\cdots,n)$ for all $t \geq 0$, we have

$\sigma(z(t)) = z(t)$ for all $t \geq 0$. Denote $D = diag\,(d_1, \cdots, d_n)$, then $d_i > 0 (i = 1, \cdots, n)$. Let $d = \min_{1 \leq i \leq n}(d_i)$. It follows from (4.16), (4.13) and Lemma 4.2 that

$$\begin{aligned}
\dot{E}(t) &= \left[z(t) - \left((x_0 - h)e^{-t} + Wz(t) + h\right)\right]^T D \\
&\quad \times \left[-z(t) + \sigma\left((x_0 - h)e^{-t} + Wz(t) + h\right)\right] \\
&= \left[z(t) - \left((x_0 - h)e^{-t} + Wz(t) + h\right)\right]^T D \\
&\quad \times \left[-\sigma(z(t)) + \sigma\left((x_0 - h)e^{-t} + Wz(t) + h\right)\right] \\
&\leq -d \cdot \|\sigma(z(t)) - \sigma\left((x_0 - h)e^{-t} + Wz(t) + h\right)\|^2 \\
&= -d \cdot \|z(t) - \sigma\left((x_0 - h)e^{-t} + Wz(t) + h\right)\|^2 \\
&= -d \cdot \|\dot{z}(t)\|^2, \qquad (4.17)
\end{aligned}$$

for all $t \geq 0$. Thus, $E(t)$ is monotonically decreasing. Since $E(t)$ is bounded, there must exist a constant $E_0$ such that

$$\lim_{t \to +\infty} E(t) = E_0 < -\infty.$$

From (4.17),

$$\begin{aligned}
\int_0^{+\infty} \|\dot{z}(s)\|^2 ds &= \lim_{t \to +\infty} \int_0^t \|\dot{z}(s)\|^2 ds \\
&\leq \lim_{t \to +\infty} \int_0^t \frac{-\dot{E}(s)}{d} ds \\
&= -\frac{1}{d} \lim_{t \to +\infty} E(t) \\
&= \frac{-E_0}{d} \\
&< +\infty.
\end{aligned}$$

Since $z(t)$ is bounded, from (4.13), it follows that $\dot{z}(t)$ is bounded. Then, $z(t)$ is uniformly continuous on $[0, +\infty)$. Again, from (4.13), it follows that $\|\dot{z}(t)\|^2$ is also uniformly continuous on $[0, +\infty)$. Thus, it must hold that

$$\lim_{t \to +\infty} \|\dot{z}(t)\|^2 = 0.$$

This completes the proof.

THEOREM 4.3 *Suppose there exists a diagonal positive definite matrix D such that DW is symmetric. If the network (4.1) is bounded, then the network (4.1) is completely stable.*

*Proof*: By Lemma 4.4, for any $x_0 \in R^n$, the solution of (4.1) starting from this point can be represented as:

$$x(t) = (x_0 - h)e^{-t} + Wz(t) + h, \tag{4.18}$$

for $t \geq 0$, where $z(t)$, which depends on $x_0$, is the solution of (4.13) starting from the origin.

Let us first show that $z(t)$ is bounded. By the assumption that $x(t)$ is bounded, there exists a constant $K > 0$ such that $\|x(t)\| \leq K$ for all $t \geq 0$. From (4.14),

$$z(t) = \int_0^t e^{-(t-s)}\sigma(x(s))ds,$$

for $t \geq 0$. Thus,

$$\begin{aligned} \|z(t)\| &\leq \int_0^t e^{-(t-s)}\|\sigma(x(s))\|ds \\ &\leq \int_0^t e^{-(t-s)}Kds \\ &\leq K, \end{aligned}$$

for $t \geq 0$. This shows that $z(t)$ is bounded.

Using Lemma 4.6, $\lim_{t\to+\infty} \dot{z}(t) = 0$. Then, from (4.18), it follows that

$$\lim_{t\to+\infty} \dot{x}(t) = W \lim_{t\to+\infty} \dot{z}(t) = 0.$$

By the definition of $\sigma(\cdot)$ and using the fact that $A$ is a symmetric matrix, $x(t)$ must be a function as:

$$x(t) = F\left(e^{\lambda_1 t}, \cdots, e^{\lambda_m t}\right) \tag{4.19}$$

where $F$ is a polynomial and $\lambda_1, \cdots, \lambda_m$ are some constants. Clearly, $\dot{x}(t) \to 0$ as $t \to +\infty$ implies that $x(t)$ is convergent, i.e., there exists a vector $x^* \in R^n$ such that

$$\lim_{t\to+\infty} x(t) = x^*.$$

Clearly, $x^*$ must be an equilibrium point. The proof is completed.

## 4. Simulation Examples

In this section, examples of simulation results will be provided to illustrate and verify the theory developed.

EXAMPLE 4.1 *Consider the two-dimensional neural network:*

$$\begin{bmatrix} \dot{x}_1(t) \\ \dot{x}_2(t) \end{bmatrix} = -\begin{bmatrix} x_1(t) \\ x_2(t) \end{bmatrix} + \begin{bmatrix} 0 & -1 \\ -0.6 & 0.4 \end{bmatrix} \begin{bmatrix} \sigma(x_1(t)) \\ \sigma(x_2(t)) \end{bmatrix} + \begin{bmatrix} 0.8 \\ 0.48 \end{bmatrix} \tag{4.20}$$

*for* $t \geq 0$.

Clearly,

$$W = \begin{bmatrix} 0 & -1 \\ -0.6 & 0.4 \end{bmatrix}.$$

Let us first use Theorem 4.1 to show the boundedness of this network. Taking $\alpha_1 = \alpha_2 = 1$, since $w_{12}^+ = w_{21}^+ = 0$, it is then easy to calculate that $\gamma_1 = 1 > 0$ and $\gamma_2 = 0.6 > 0$. Thus the conditions of Theorem 4.1 are fully satisfied. Using Theorem 4.1, the network is bounded and there exists a compact set to attract the network. By simple calculations, the global attractive set is given by

$$S = \left\{ \begin{bmatrix} x_1 \\ x_2 \end{bmatrix} \middle| |x_1| \leq 1.6, |x_2| \leq 1.28 \right\}.$$

Define a diagonal positive definite matrix

$$D = \begin{bmatrix} 1 & 0 \\ 0 & \frac{5}{3} \end{bmatrix},$$

then,

$$DW = \begin{bmatrix} 0 & -1 \\ -1 & -\frac{2}{3} \end{bmatrix},$$

which is symmetric. By Theorem 4.3, this network is completely stable. Thus, every trajectory of (4.20) must converge to an equilibrium point.

The equilibrium set of the network must be located in the attractive set $S$. In fact, it is not difficult to calculate out the equilibrium set of (4.20) as

$$S^e = \left\{ \begin{bmatrix} x_1 \\ x_2 \end{bmatrix} \middle| x_1 + x_2 = 0.8, x_1 \geq 0, x_2 \geq 0 \right\}.$$

Clearly, $S^e \subset S$.

The boundedness and global attractivity of (4.20) can also be checked by Theorem 4.2. Invoking Theorem 4.2, it can calculate out another global attractive set

$$\bar{S} = \{x | \|x\| \leq 2.6077\}.$$

This also shows that Theorem 4.2 does not imply monostability.

Figure 4.2 shows the simulation results for global attractivity and complete stability of the network (4.20) for 250 trajectories originating from randomly selected initial points. It can be observed that every trajectory converges to an equilibrium point. Clearly, this network has a line attractor $S^e$. The square contained part in Figure 4.2 is the global attractive set $S$ calculated by using Theorem 4.1. The circle contained part is the global attractive set $\bar{S}$ calculated by using Theorem 4.2.

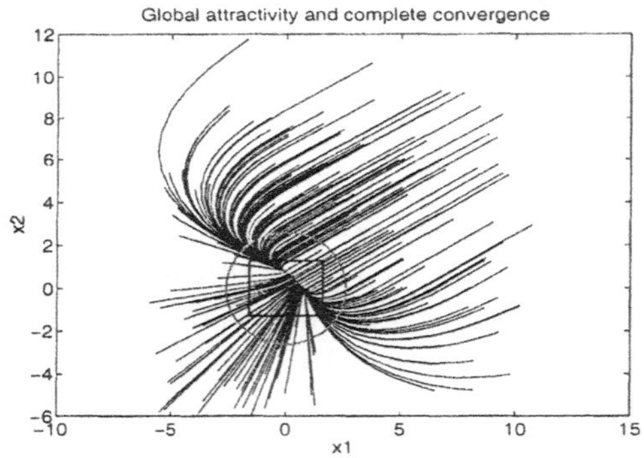


*Figure 4.2.* Global attractivity and complete convergence of network (4.20)

EXAMPLE 4.2 *Consider the following three-dimensional neural network:*

$$\dot{x}(t) = -x(t) + W\sigma(x(t)) + h, \quad (4.21)$$

*for* $t \geq 0$, *where* $x = (x_1, x_2, x_3)^T$, $h = (-0.2, -0.8, 0.3)^T$ *and*

$$W = \begin{bmatrix} 0 & 1 & 0.5 \\ -3 & 0 & 2 \\ 0.5 & -4 & 0 \end{bmatrix}.$$

It is not difficult to show that there does not exist any diagonal matrix $D$ such that $DW$ becomes symmetric. Hence, Theorem 4.3 cannot be used to ascertain whether the network is completely stable or not. However, Theorem 4.2 can be used to show the boundedness and global attractivity of the network.

Clearly,

$$I - \left[\frac{W^T + W}{2}\right]^+ = \begin{bmatrix} 1 & 0 & -0.5 \\ 0 & 1 & 0 \\ -0.5 & 0 & 1 \end{bmatrix}.$$

Its eigenvalues are 0.5, 1 and 1.5, respectively. Thus, it is a positive definite matrix. By Theorem 4.2, the network is bounded and a global attractive set can be located as

$$S = \{x \,|\, \|x\| \leq 3.3528\}.$$

Figure 4.3 shows the simulation results in three-dimensional space for 50 trajectories originating from randomly selected initial points.

Figures 4.4, 4.5 and 4.6 show the projections of the trajectories onto the phase planes $(x_1, x_2)$, $(x_1, x_3)$ and $(x_2, x_3)$, respectively. The parts contained in the circles in Figure 4.4, 4.5 and 4.6, are the projections of the global attractive set $S$ onto the corresponding phase planes.

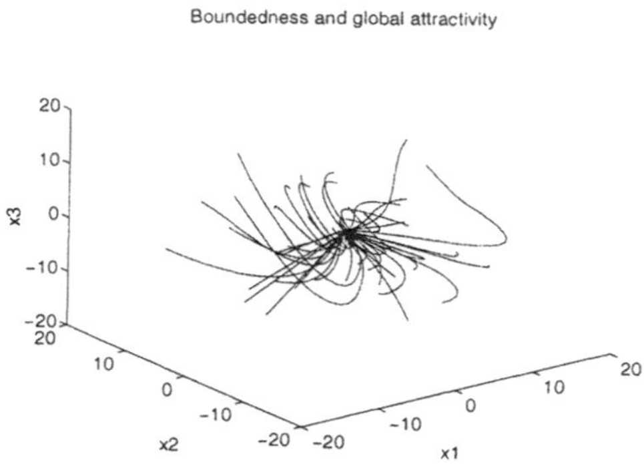


*Figure 4.3.* Boundedness and global attractivity of network (4.21)

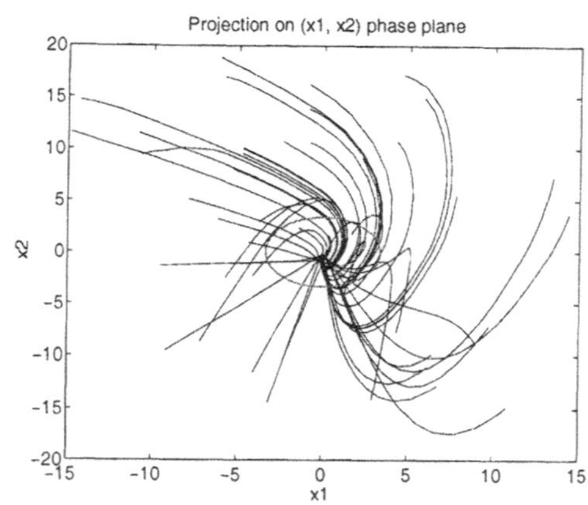


*Figure 4.4.* Projection of trajectories of (4.21) on $(x_1, x_2)$ phase plane

The boundedness of this network cannot be checked by Theorem 4.1. In fact,

$$I - W^+ = \begin{bmatrix} 1 & -1 & -0.5 \\ 0 & 1 & -2 \\ -0.5 & 0 & 1 \end{bmatrix},$$

which has a negative eigenvalue which obviously violates the conditions of Theorem 4.1.

From the simulation results above, it can be observed that this network seems to have monostability property. However, this point cannot be confirmed at present. It is worthwhile noting that this network does not satisfy the exist-

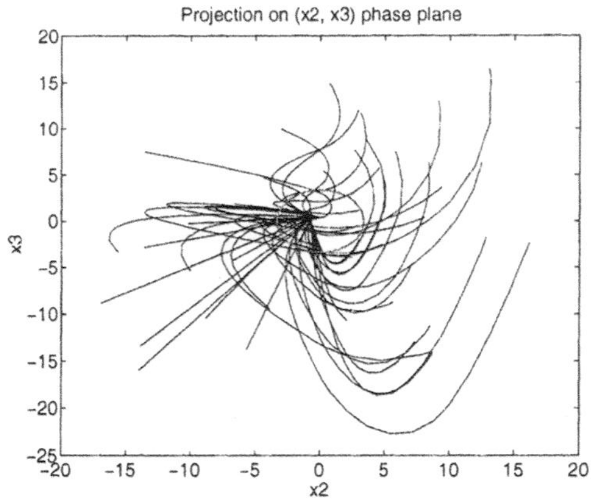


*Figure 4.5.* Projection of trajectories of (4.21) on $(x_2, x_3)$ phase plane

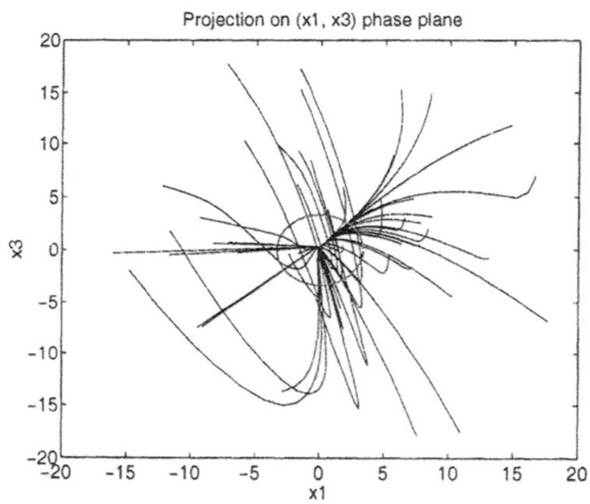


*Figure 4.6.* Projection trajectories of (4.21) on $(x_1, x_3)$ phase plane

ing stability conditions, for examples the global stability conditions recently reported in [57, 111, 220, 218, 219].

Apart from the presented simulations, the boundedness, global attractivity and complete stability of networks that belong to the network model (4.1) are extensively simulated . All the results verify the theory developed. In fact, certain global attractive sets approach the critical bounds containing the equilibrium points.

## 5. Conclusions

In this chapter, multistability of recurrent neural networks with unsaturating piecewise linear transfer functions has been studied. Three basic dynamical problems are addressed for this class of networks: boundedness, global attractivity and complete stability. Local inhibition based conditions for boundedness and global attractivity are established. Under these conditions, the global attractive sets of the networks are obtained. With symmetry conditions attached to the networks weights matrix, the complete stability is analyzed by showing each trajectory converges to an equilibrium point.

This model is under going research both in theory and applications. It has quite simple structure and is easily to be implemented in hardware. It is believed that more interesting results on this model will come out. For some recent important works, refer to [208].

Chapter 5

# LOTKA-VOLTERRA RECURRENT NEURAL NETWORKS WITH DELAYS

## 1. Introduction

The Lotka-Volterra model of RNNs was first proposed in [62]. Derived from conventional membrane dynamics of competing neurons, it has found successful applications in winner-take-all, winner-share-all and $k$-winner-take-all problems, see [15, 16, 62]. Due to the application potential of this class of neural networks, it is necessary and useful to study its general dynamic properties.

In this chapter, convergence of general Lotka-Volterra RNNs with delays will be studied. The model is described by the following nonlinear differential equation with delays

$$\dot{x}_i(t) = x_i(t)\left[h_i - x_i(t) + \sum_{j=1}^{n}\Big(a_{ij}x_j(t) + b_{ij}x_j(t-\tau_{ij}(t))\Big)\right] \tag{5.1}$$

for $t \geq 0$ and $(i = 1, \cdots, n)$, where each $x_i$ denotes the activity of neuron $i$, $A = (a_{ij})_{n\times n}$ and $B = (b_{ij})_{n\times n}$ are real $n \times n$ matrices, each of their elements denotes the synaptic weights and represents the strength of the synaptic connection from neuron $j$ to neuron $i$, $x = (x_1, \cdots, x_n)^T \in R^n$, $h_i \in R^n$ denotes external inputs, the delays $\tau_{ij}(t)(i, j = 1, \cdots, n)$ are non-negative continuous functions satisfying $0 \leq \tau_{ij}(t) \leq \tau$ for all $t \geq 0$, where $\tau \geq 0$ is a constant.

Incorporating delays in neural network models is important both in theory and applications. In [85], Hopfield realized that in hardware implementations, time delays occur due to the finite switching speed of the amplifiers. It is known that delays can affect the dynamic behavior of neural networks [17, 137]. A stable network may become unstable by adding some delays [67]. Delayed neural networks have found many applications in the processing of moving images, image compression [191], etc. In recent years, neural networks with delays have been widely studied, see for examples, [8, 17, 52, 67, 69, 93, 127, 128, 137, 149, 159, 179, 211, 191, 218, 219]. Today, delays have been widely accepted as important parameters associated with neural network models.

Both multistability and monostability of the network (5.1) will be studied in this chapter. In section 2, multistability is studied by using local inhibition method. Three important properties of the network (5.1): boundedness, global attractivity and complete convergence will be addressed.

In section 3, monostability of network (5.1) will be studied. Monostability of recurrent neural networks is an important dynamic property of neural networks for certain applications, including those relating to optimization. The monostability of neural networks has been widely studied in recent years.

Three dynamic properties of the network (5.1): lower positive boundedness, global exponential convergence and global asymptotic stability will be addressed.

## 2. Multistability Analysis

This section study the multistability of network (5.1) by using the method of local inhibition. Preliminaries will be given in subsection 2.1. Boundedness and global attractivity will be discussed in subsection 2.2. In subsection 2.3, complete stability will be discussed. Simulations and examples will be given in subsection 2.4.

### 2.1 Preliminaries

DEFINITION 5.1 *The network (5.1) is said to be bounded if each of its trajectories is bounded.*

DEFINITION 5.2 *Let $S$ be a compact subset of $R^n$. Denote the $\epsilon-$neighborhood of $S$ by $S_\epsilon$. The compact set $S$ is said to globally attract the network (5.1), if for any $\epsilon > 0$, all trajectories of (5.1) ultimately enter and remain in $S_\epsilon$.*

A bounded network does not imply it will have global attractive sets. For example, consider a simple system $\dot{x}(t) = 0$ for $t \geq 0$. Clearly, it is bounded but there does not exist any compact set to globally attract its trajectories. However, if a network possesses global attractive sets, then it must be bounded.

Denote

$$R^n_+ = \{x | x \in R^n, x_i \geq 0, (i = 1, \cdots, n)\}.$$

DEFINITION 5.3 *A vector $x^* = (x_1^*, \cdots, x_n^*)^T \in R^n_+$ is called an equilibrium point of (5.1) in $R^n_+$, if*

$$x_i^* \left[ h_i - x_i^* + \sum_{j=1}^{n} (a_{ij} + b_{ij}) x_j^* \right] \equiv 0, (i = 1, \cdots, n).$$

Denote by $S^e$ the set contains all the equilibrium of (5.1).

In this section, the equilibrium points which are located in $R^n_+$ will be considered.

Assume the initial condition as

$$\begin{cases} x_i(t) = \phi_i(t) \geq 0, & t \in [-\tau, 0], \\ \phi_i(0) > 0, & (i = 1, \cdots, n), \end{cases} \tag{5.2}$$

where each $\phi_i$ is a continuous function defined on $[-\tau, 0]$.

DEFINITION 5.4 *The network (5.1) is said to be almost completely stable, if the equilibrium set $S^e$ is not empty and every trajectory $x(t)$ of (5.1) converges to $S^e$, i.e.*

$$dist\,(x(t), S^e) \triangleq \min_{x^* \in S^e} \|x(t) - x^*\| \to 0$$

*as $t \to +\infty$.*

LEMMA 5.1 *Each solution $x(t)$ of (5.1) with the initial condition (5.2) satisfies*

$$x_i(t) > 0, (i = 1, \cdots, n),$$

*for all $t \geq 0$.*

*Proof:* Denote

$$r_i(t) = h_i - x_i(t) + \sum_{j=1}^{n} a_{ij} x_j(t) + \sum_{j=1}^{n} b_{ij} x_j(t - \tau_{ij}(t)),$$

for $t \geq 0$ and $(i = 1, \cdots, n)$. Then, from (5.1), it follows that

$$x_i(t) = \phi_i(0) e^{\int_0^t r_i(s) ds} > 0, (i = 1, \cdots, n)$$

for $t \geq 0$. This completes the proof.

Lemma 5.1 shows an interesting property; if the initial condition satisfies (5.1), then the corresponding trajectory stays in positive domain of $R^n$. This property will allow in the next section to use local inhibitions to guarantee boundedness of the network (5.1).

Throughout this section, for any constant $c \in R$, denote

$$c^+ = \max(0, c).$$

A continuous function $f(t)$ defined on $[0, +\infty)$ is said to be in $L^2$ space, if

$$\int_0^{+\infty} f^2(s) ds < +\infty.$$

## 2.2 Boundedness and Global Attractivity

The trajectories of the Lotka-Volterra neural network (5.1) may diverge. For example, consider the simple one dimensional network

$$\dot{x}(t) = x(t)\left[1 + 2x(t)\right],$$

for $t \geq 0$. Given any $x(0) > 0$, it follows that

$$\frac{x(t)}{1+2x(t)} = \frac{x(0)}{1+2x(0)} e^{t},$$

for $t \geq 0$. Taking $x(0) = 1$ in particular, it is easy to see that $x(t)$ diverges at $t = \ln(3/2)$.

Boundedness is a basic enabling property for recurrent neural networks in practical applications. It is necessary to derive conditions for boundedness of the network (5.1). The classical method to obtain conditions of boundedness is to restrict the weights to be sufficiently small, see for examples [220, 219]. However, these conditions imply global convergence, that is all the trajectories of a network converge to an equilibrium point. Thus, the network is actually monostable. In [75], it is pointed out that monostable networks are computationally restrictive: they cannot deal with important neural computations such as those governing the decision making process. There are many applications in which a multi-stability property is a necessary condition. In [75, 203, 223], multistability for neural networks with unsaturating piecewise linear transfer functions are studied. The boundedness conditions that we will present in the following allow for such multistable behavior. This will be achieved by applying local inhibition to the network's weights.

THEOREM 5.1 *If there exist constants $\gamma_i > 0 (i = 1, \cdots, n)$ such that*

$$\beta_i \triangleq 1 - a_{ii} - \frac{1}{\gamma_i}\sum_{j=1}^{n}\gamma_j\left[a_{ij}^{+}\cdot(1-\delta_{ij}) + b_{ij}^{+}\right] > 0, \quad (i = 1, \cdots, n),$$

*where*

$$\delta_{ij} = \begin{cases} 1, & i = j \\ 0, & i \neq j \end{cases},$$

*then the network (5.1) is bounded. Moreover, the compact set*

$$S = \left\{ x \,\middle|\, 0 \leq x_i \leq \gamma_i \cdot \max_{1\leq j\leq n}\left\{\frac{h_j}{\beta_j}, 0\right\}, (i = 1, \cdots, n)\right\}$$

*globally attracts the network (5.1).*

*Proof*: Firstly, it will show that the network (5.1) is bounded. Clearly,

$$x_i(t)\begin{cases} \geq 0, & -\tau \leq t < 0 \\ > 0, & t \geq 0, \end{cases} \qquad (i = 1, \cdots, n).$$

Then, from (5.1),

$$\begin{aligned} \dot{x}_i(t) \leq\; & x_i(t)\Bigg[h_i - (1 - a_{ii})x_i(t) \\ & + \sum_{j=1}^{n}\left(a_{ij}^{+}\cdot(1-\delta_{ij})\,x_j(t) + b_{ij}^{+}x_j(t-\tau_{ij}(t))\right)\Bigg], \end{aligned} \quad (5.3)$$

for $t \geq 0$ and $(i = 1, \cdots, n)$.

Define

$$z_i(t) = \frac{x_i(t)}{\gamma_i}, (i = 1, \cdots, n),$$

for $t \geq -\tau$. Obviously,

$$z_i(t)\begin{cases} \geq 0, & -\tau \leq t < 0 \\ > 0, & t \geq 0, \end{cases} \qquad (i = 1, \cdots, n).$$

Moreover, from (5.3), it follows that

$$\begin{aligned} \dot{z}_i(t) \leq\; & x_i(t)\Bigg[h_i - (1 - a_{ii})z_i(t) \\ & + \frac{1}{\gamma_i}\sum_{j=1}^{n}\gamma_j\left(a_{ij}^{+}\cdot(1-\delta_{ij})\,z_j(t) + b_{ij}^{+}z_j(t-\tau_{ij}(t))\right)\Bigg], \end{aligned} \quad (5.4)$$

for $t \geq 0$.

Denote

$$|\phi| = \max_{1\leq i\leq n}\left[\sup_{-\tau\leq s\leq 0}\phi_i(s)\right],$$

and

$$\Pi = \max_{1\leq i\leq n}\left\{\frac{1}{\gamma_i}\right\}\cdot\max_{1\leq i\leq n}\left\{|\phi| + 1, \frac{h_i}{\beta_i}\right\}.$$

It will be proven that

$$z_i(t) < \Pi, (i = 1, \cdots, n), \quad (5.5)$$

for all $t \geq 0$. Otherwise, if (5.5) is not true, since $z_i(t) \leq |\phi|/\gamma_i < \Pi(i = 1, \cdots, n)$ for $t \in [-\tau, 0]$, there must exist a $t_1 > 0$ such that $\dot{z}_i(t_1) \geq 0$ and

$$z_i(t_1) = \Pi; \quad z_j(t)\begin{cases} < \Pi, & -\tau \leq t < t_1, \quad i = j \\ \leq \Pi, & -\tau \leq t \leq t_1, \quad i \neq j \end{cases}.$$

However, from (5.4),

$$\begin{aligned}\dot{z}_i(t_1) &\leq x_i(t_1)\left[h_i - (1-a_{ij})\Pi + \frac{1}{\gamma_i}\sum_{j=1}^{n}\gamma_j\left(a_{ij}^{+}\cdot(1-\delta_{ij})\Pi + b_{ij}^{+}\Pi\right)\right] \\ &= x_i(t_1)\cdot(h_i - \beta_i\cdot\Pi) \\ &< 0.\end{aligned}$$

This poses a contradiction and it implies that (5.5) is true. Thus, the network (5.1) is bounded.

Next, the global attractivity of the network will be proven. Denote

$$m = \max_{1\leq i\leq n}\left\{\frac{h_i}{\beta_i}, 0\right\}.$$

Given any $\epsilon > 0$, clearly, a constant $\eta > 0$ can be chosen such that

$$\frac{1 - a_{ii} - \beta_i}{\beta_i}\cdot\eta \leq \frac{\epsilon}{2}, (i = 1, \cdots, n). \tag{5.6}$$

Let $N$ be the first non-negative integer such that $m + \epsilon + N\cdot\eta \geq \Pi$.

Define

$$t_k = kT^*, k = 1, 2, \cdots, N,$$

where

$$T^* = \tau + \frac{2}{\beta\gamma\epsilon}\cdot\ln\left(\frac{2\Pi}{m+\epsilon}\right), \quad \beta = \min_{1\leq i\leq n}\{\beta_i\}, \quad \gamma = \min_{1\leq i\leq n}\{\gamma_i\}.$$

We will prove that

$$z_i(t) \leq m + \epsilon + (N-k)\eta, \quad t \geq t_k, \tag{5.7}$$

for $k = 0, 1, 2, \cdots, N$ by mathematical induction.

By (5.5), clearly, (5.7) holds for $k = 0$. Suppose (5.7) holds for some $k (0 \leq k < N)$, i.e.,

$$z_i(t) \leq m + \epsilon + (N-k)\eta, (i = 1, \cdots, n), \tag{5.8}$$

for $t \geq t_k$. It will prove that

$$z_i(t) \leq m + \epsilon + (N-k-1)\eta, (i = 1, \cdots, n), \tag{5.9}$$

for $t \geq t_{k+1}$.

Two steps will be used to prove (5.9). In the first step, it will be proven that there exists a $\bar{t} \in [t_k + \tau, t_{k+1}]$ such that

$$z_i(\bar{t}) \leq m + \epsilon + (N-k-1)\eta, (i = 1, \cdots, n). \tag{5.10}$$

Suppose (5.10) is not true, then there must exist some $i$ such that

$$z_i(t) > m + \epsilon + (N - k - 1)\eta > m + \epsilon, \quad (5.11)$$

for all $t \in [t_k + \tau, t_{k+1}]$. From (5.8) and (5.11), it follows that

$$\sup_{t-\tau \le \theta \le t} z_j(\theta) \le m + \epsilon + (N - k)\eta < z_i(t) + \eta, (j = 1, \cdots, n), \quad (5.12)$$

for $t \in [t_k + \tau, t_{k+1}]$. Then, for $t \in [t_k + \tau, t_{k+1}]$, it follows from (5.3), (5.12), (5.11) and (5.6) that

$$\begin{aligned}
\frac{d \ln z_i(t)}{dt} &\le \gamma_i \Bigg[ h_i - (1 - a_{ii}) z_i(t) \\
&\quad + \sum_{j=1}^{n} \left( a_{ij}^{+} \cdot (1 - \delta_{ij})\, z_j(t) + b_{ij}^{+} z_j(t - \tau_{ij}(t)) \right) \Bigg] \\
&\le \gamma_i \Bigg[ h_i - (1 - a_{ii}) z_i(t) \\
&\quad + \sum_{j=1}^{n} \left( a_{ij}^{+} \cdot (1 - \delta_{ij}) + b_{ij}^{+} \right) (z_i(t) + \eta) \Bigg] \\
&\le \gamma_i \Big[ h_i - (1 - a_{ii}) \cdot z_i(t) + (1 - a_{ii} - \beta_i)(z_i(t) + \eta) \Big] \\
&\le \gamma_i \left[ h_i - \beta_i z_i(t) + (1 - a_{ii} - \beta_i)\, \eta \right] \\
&\le \gamma_i \beta_i \left[ \frac{h_i}{\beta_i} - (m + \epsilon) + \frac{1 - a_{ii} - \beta_i}{\beta_i} \eta \right] \\
&\le \gamma_i \beta_i \left[ \frac{h_i}{\beta_i} - (m + \epsilon) + \frac{\epsilon}{2} \right] \\
&\le -\frac{\epsilon \beta_i \gamma_i}{2} \\
&\le -\frac{\epsilon \beta \gamma}{2}.
\end{aligned}$$

Thus,

$$\ln z_i(t_{k+1}) - \ln z_i(t_k + \tau) \le -\frac{\epsilon \beta \gamma}{2} \cdot (T^* - \tau),$$

and so

$$\begin{aligned}
z_i(t_{k+1}) &\le z_i(t_k + \tau) e^{-\frac{\epsilon \beta \gamma}{2} \cdot (T^* - \tau)} \\
&\le \Pi \cdot e^{-\ln\left(\frac{2\Pi}{m+\epsilon}\right)} \\
&= \frac{m + \epsilon}{2}.
\end{aligned}$$

This contradicts (5.11). Therefore, (5.10) is true.

In the second step, it will prove that

$$z_i(t) \le m + \epsilon + (N - k - 1)\eta, (i = 1, \cdots, n), \tag{5.13}$$

for all $t \ge \bar{t}$. Otherwise, there must exist a $\hat{t} > \bar{t}$ and some $i$ such that

$$z_i(\hat{t}) > m + \epsilon + (N - k - 1)\eta, \quad \dot{z}_i(\hat{t}) > 0. \tag{5.14}$$

From (5.8) and (5.14),

$$z_j(\theta) \le m + \epsilon + (N - k)\eta < z_i(\hat{t}) + \eta, (j = 1, \cdots, n), \tag{5.15}$$

for $\theta \ge t_k$. Thus, from (5.3), (5.15) and (5.6), it follows that

$$\begin{aligned}
\dot{z}_i(\hat{t}) &\le x_i(\hat{t})\Bigg[h_i - (1 - a_{ii})z_i(\hat{t}) \\
&\quad + \sum_{j=1}^{n} \left(a_{ij}^+ \cdot (1 - \delta_{ij})\, z_j(\hat{t}) + b_{ij}^+ z_j(\hat{t} - \tau_{ij}(\hat{t}))\right)\Bigg] \\
&\le x_i(\hat{t})\Bigg[h_i - (1 - a_{ii})z_i(\hat{t}) \\
&\quad + \sum_{j=1}^{n} \left(a_{ij}^+ \cdot (1 - \delta_{ij}) + b_{ij}^+\right) \left(z_i(\hat{t}) + \eta\right)\Bigg] \\
&\le x_i(\hat{t})\left[h_i - (1 - a_{ii}) \cdot z_i(\hat{t}) + (1 - a_{ii} - \beta_i)\left(z_i(\hat{t}) + \eta\right)\right] \\
&\le x_i(\hat{t})\left[h_i - \beta_i z_i(\hat{t}) + (1 - a_{ii} - \beta_i)\,\eta\right] \\
&= x_i(\hat{t})\beta_i \left[\frac{h_i}{\beta_i} - (m + \epsilon) + \frac{1 - a_{ii} - \beta_i}{\beta_i}\eta\right] \\
&\le -x_i(\hat{t})\frac{\epsilon\beta_i}{2} \\
&< 0.
\end{aligned}$$

which contradict (5.14). This proves that (5.13) is true.

Since $t_{k+1} \ge \bar{t}$, then using (5.13), it follows that (5.9) holds. By mathematical induction, (5.7) must be true. This completes the proof of (5.7).

Taking $k = N$ in (5.7), then

$$z_i(t) \le m + \epsilon, (i = 1, \cdots, n),$$

for all $t \ge t_N$. Thus, it follows that

$$x_i(t) \le \gamma_i \cdot (m + \epsilon), (i = 1, \cdots, n),$$

for all $t \geq 0$. This shows that the set $S$ globally attracts the trajectories of (5.1). This completes the proof.

The conditions for boundedness given in Theorem 5.1 use the local inhibition of the network (5.1), i.e., it only use part of the weights of the network to achieve non-divergence. Due to this fact, the network is allowed to have the property of multistability. Examples in Subsection 2.4 will further confirm this point. It is interesting that following Theorem 5.1, the global attractive compact set can be calculated explicitly.

Taking $\gamma_i = 1 (i = 1, \cdots, n)$ in Theorem 5.1, it gives an especially simple results.

COROLLARY 5.1 *If*

$$\bar{\beta}_i \triangleq 1 - a_{ii} - \sum_{j=1}^{n} \left[ a_{ij}^{+} \cdot (1 - \delta_{ij}) + b_{ij}^{+} \right] > 0, \quad (i = 1, \cdots, n)$$

*where*

$$\delta_{ij} = \begin{cases} 1, & i = j \\ 0, & i \neq j \end{cases},$$

*then the network (5.1) is bounded. Moreover, the compact set*

$$S = \left\{ x \middle| 0 \leq x_i \leq \max_{1 \leq j \leq n} \left\{ \frac{h_j}{\bar{\beta}_j}, 0 \right\}, (i = 1, \cdots, n) \right\},$$

*globally attracts the network (5.1).*

## 2.3 Complete Convergence

In certain applications, a network which possess the non-divergence property is not sufficient. More desirable dynamic properties are required to enable the network to be effectively used. Convergence is one of the most important properties of a recurrent neural network towards applications. In this section, the complete stability of the network (5.1) will be analyzed. This property guarantees a network to work well without exhibiting any oscillations or chaotic behavior. In addition, a complete stable network may possess multistability property. Stable and unstable equilibrium points may co-exist in a completely stable network. This property has its important applications in certain networks [203]. Complete stability analysis for other models of neural networks could be found in [67, 179, 85, 205].

THEOREM 5.2 *Suppose the network (5.1) is bounded. If there exists a diagonal matrix $D$ with positive elements such that the matrix $D(A+B)$ is symmetric, and*

$$\tau_{ij}(t) \in L^2, \quad (i, j = 1, \cdots, n), \qquad (5.16)$$

*then the network (5.1) is almost completely stable.*

*Proof:* Since the network is bounded, there exists a constant $c > 0$ such that

$$0 < x_i(t) \leq c, \quad (i = 1, \cdots, n),$$

for $t \geq 0$. From (5.1), clearly, $\dot{x}_i(t)(i = 1, \cdots, n)$ are bounded. Then, there exists a constant $m > 0$ such that

$$|\dot{x}_i(t)| \leq m, \quad (i = 1, \cdots, n), \tag{5.17}$$

for $t \geq 0$.

Constructing an energy function

$$E(t) = -H^T Dx(t) + \frac{1}{2}x^T(t)D(I - A - B)x(t). \tag{5.18}$$

for $t \geq 0$, where $I$ is the $n \times n$ identity matrix, and $H = (h_1, \cdots, h_n)^T$. Suppose $D = diag(d_1, \cdots, d_n)$, clearly, $d_i > 0(i = 1, \cdots, n)$. Using the fact that $D(A + B)$ is symmetric, from (5.18) and (5.1), it follows that

$$\begin{aligned}
\dot{E}(t) &= -\sum_{i=1}^{n} d_i \left( h_i - x_i(t) + \sum_{j=1}^{n} (a_{ij} + b_{ij}) x_j(t) \right) \dot{x}_i(t) \\
&= -\sum_{i=1}^{n} d_i \left( h_i - x_i(t) + \sum_{j=1}^{n} \left( a_{ij} x_j(t) + b_{ij} x_j(t - \tau_{ij}(t)) \right) \right) \dot{x}_i(t) \\
&\quad + \sum_{i=1}^{n} \sum_{j=1}^{n} d_i b_{ij} \left( \int_t^{t-\tau_{ij}(t)} \dot{x}_i(s) ds \right) \dot{x}_i(t) \\
&= -\sum_{i=1}^{n} d_i \frac{\dot{x}_i^2(t)}{x_i(t)} + \sum_{i=1}^{n} \sum_{j=1}^{n} d_i b_{ij} \left( \int_t^{t-\tau_{ij}(t)} \dot{x}_i(s) ds \right) \dot{x}_i(t) \\
&\leq -\sum_{i=1}^{n} d_i \frac{\dot{x}_i^2(t)}{x_i(t)} + \sum_{i=1}^{n} d_i \sum_{j=1}^{n} |b_{ij}| \left( \int_{t-\tau_{ij}(t)}^{t} |\dot{x}_i(s)| ds \right) |\dot{x}_i(t)| \\
&\leq -\sum_{i=1}^{n} \frac{d_i}{c} \dot{x}_i^2(t) + \sum_{i=1}^{n} d_i \sum_{j=1}^{n} |b_{ij}| m \tau_{ij}(t) |\dot{x}_i(t)| \\
&= -\sum_{i=1}^{n} \frac{d_i}{c} \dot{x}_i^2(t) + \sum_{i=1}^{n} d_i \sum_{j=1}^{n} \left( \frac{b_{ij}^2 m^2 nc}{2} \tau_{ij}^2(t) + \frac{\dot{x}_i^2(t)}{2nc} \right) \\
&= -\sum_{i=1}^{n} \frac{d_i}{2c} \dot{x}_i^2(t) + \sum_{i=1}^{n} \sum_{j=1}^{n} \frac{d_i n c b_{ij}^2 m^2}{2} \tau_{ij}^2(t) \\
&\leq -\delta \|\dot{x}(t)\|^2 + \eta \sum_{i=1}^{n} \sum_{j=1}^{n} \tau_{ij}^2(t),
\end{aligned}$$

for $t \geq 0$, where

$$\delta = \min_{1 \leq i \leq n} \left\{ \frac{d_i}{2c} \right\}, \quad \eta = \max_{1 \leq i,j \leq n} \left\{ \frac{d_i n c b_{ij}^2 m^2}{2} \right\}.$$

Denote

$$\tilde{E}(t) = E(t) - \eta \sum_{i=1}^{n} \sum_{j=1}^{n} \int_0^t \tau_{ij}^2(s) ds, \quad t \geq 0.$$

Then,

$$\dot{\tilde{E}}(t) \leq -\delta \|\dot{x}(t)\|^2, \tag{5.19}$$

for $t \geq 0$. Clearly, $\tilde{E}(t)$ is monotone decreasing. Since $E(t)$ is bounded and following (5.16),

$$\int_0^{+\infty} \tau_{ij}^2(s) ds < +\infty, (i, j = 1, \cdots, n)$$

then, $\tilde{E}(t)$ must be bounded. Thus, there must exist a constant $\tilde{E}_0$ such that

$$\lim_{t \to +\infty} \tilde{E}(t) = \tilde{E}_0.$$

From (5.19),

$$\|\dot{x}(t)\|^2 \leq -\frac{1}{\delta} \cdot \dot{\tilde{E}}(t),$$

for $t \geq 0$. Then,

$$\begin{aligned} \int_0^{+\infty} \|\dot{x}(s)\|^2 ds &\leq \frac{1}{\delta} \cdot \left( \tilde{E}(0) - \lim_{t \to +\infty} \tilde{E}(t) \right) \\ &= \frac{1}{\delta} \cdot \left( E(0) - \tilde{E}_0 \right) \\ &< +\infty. \end{aligned}$$

From (5.17), $\dot{x}_i(t) (i = 1, \cdots, n)$ are bounded, then each $x_i(t)$ is uniformly continuous on $[-\tau, +\infty)$. From (5.1), it follows that each $\dot{x}_i(t)$ and then $\|\dot{x}(t)\|^2$ is uniformly continuous on $[0, +\infty)$. Thus, it must have

$$\lim_{t \to +\infty} \|\dot{x}(t)\|^2 = 0,$$

and so

$$\lim_{t \to +\infty} \dot{x}_i(t) = 0, (i = 1, \cdots, n).$$

Since $x(t)$ is bounded, every subsequence of $\{x(t)\}$ must contain convergent subsequence. Let $\{x(t_m)\}$ be any of such a convergent subsequence. There exists a $x^* \in R^n$ such that

$$\lim_{t_m \to +\infty} x(t_m) = x^*.$$

Then, from (5.1)

$$-x^* + W\sigma(x^*) + h = \lim_{t_m \to +\infty} \dot{x}(t_m) = 0.$$

Thus, $x^* \in S^e$. This shows that $S^e$ is not empty and any convergent subsequence of $\{x(t)\}$ converges to a point of $S^e$.

Next, the method in [114] will be used to prove

$$\lim_{t \to +\infty} dist\,(x(t), S^e) = 0. \tag{5.20}$$

Suppose (5.20) is not true, then, there exists a constant $\epsilon_0 > 0$ such that for any $T \geq 0$, there exists a $\bar{t} \geq T$ satisfies $dist\,(x(\bar{t}), S^e) \geq \epsilon_0$. From this and by the boundedness property of $x(t)$, it can choose a convergent subsequence $\{x(\bar{t}_m)\}$, where $\lim_{\bar{t}_m \to +\infty} x(\bar{t}_m) = x^\dagger \in S^e$, such that

$$dist\,(x(\bar{t}_m), S^e) \geq \epsilon_0, (m = 1, 2, \cdots).$$

Letting $t_m \to +\infty$,

$$dist\left(x^\dagger, S^e\right) \geq \epsilon_0 > 0$$

which contradict to $dist\left(x^\dagger, S^e\right) = 0$ since $x^\dagger \in S^e$. This proves (5.20) is true. Then, the network (5.1) is almost completely stable. The proof is completed.

The condition (5.16) requires that the delays are in the $L^2$ space. Thus, the delays will decay as time approaches infinity. Decaying of delays in neural networks seems reasonable, since delays occur usually at the beginning of the operations of networks due to the hardware characteristics such as switching delays, parameter variability, parasitic capacitance, and inductance etc.

Taking $D = I$ in Theorem 5.2, it gives a following simple result.

COROLLARY 5.2 *Suppose the network (5.1) is bounded. If $(A + B)$ is a symmetrical matrix, and*

$$\tau_{ij}(t) \in L^2, \quad (i, j = 1, \cdots, n)$$

*then the network (5.1) is almost completely stable.*

It is guessed that the almost complete stability results of this section could be further improved to complete stability, i.e., each trajectory converges to an equilibrium point instead of converges to the equilibrium set. However, it could not be proven this at present. It seems this problem contains some unexpected difficulties. It would be an interesting problem to study in the future.

## 2.4 Simulations and Examples

EXAMPLE 5.1 *Consider the two-dimensional network*

$$\begin{cases} \dot{x}_1(t) = x_1(t) \cdot \Big(3 - x_1(t) + 0.5x_1(t) - x_2(t - \tau_1(t))\Big) \\ \\ \dot{x}_2(t) = x_2(t) \cdot \Big(3 - x_2(t) - x_1(t - \tau_2(t)) + 0.5x_2(t)\Big) \end{cases} \tag{5.21}$$

*for* $t \geq 0$, *where* $\tau_i(t)(i = 1, 2)$ *are bounded non-negative continuous functions.*

It is easy to check that the conditions of Theorem 5.1 are satisfied. Thus, it is bounded. By Theorem 5.1, it can be calculated that there is a compact set

$$S = \{0 \leq x_1 \leq 6; \quad 0 \leq x_2 \leq 6\},$$

which globally attracts all the trajectories of the network. Clearly, all the equilibrium points in $R_+^2$ of the network must be located in the set $S$. In fact, it can be calculated out that this network has four equilibrium points in $R_+^2$. They are $(0, 0)$, $(6, 0)$, $(0, 6)$ and $(2, 2)$. Clearly, they are all located in $S$. If $\tau_i(t) \in L^2(i = 1, 2)$, say $\tau_i(t) = e^{-t}(i = 1, 2)$, then the conditions of Theorem 5.2 are satisfied and the network (5.21) is completely convergent.

Figure 5.1 shows the simulation results of the global attractivity and complete convergence of the network. For the convenience of simulation, we assume $\tau_i(t) \equiv 0(i = 1, 2)$. The part contained in the square shown in the figure is the set $S$. It contains all the equilibrium points of the network (5.21). Among these equilibrium points, $(6, 0)$ and $(0, 6)$ are stable, while $(0, 0)$ and $(2, 2)$ are unstable. This example also shows that in a completely stable neural network, stable and unstable equilibrium points can co-exist in the network. Existing unstable directions in the state space of a neural networks is essential for certain neural computations [203].

EXAMPLE 5.2 *Consider the following ten-dimensional Lotka-Volterra neural network with one delay,*

$$\dot{x}_i(t) = x_i(t) \left[ h_i - x_i(t) - \sum_{j=1, j \neq i}^{n} \Big(x_j(t) + x_j(t - \tau(t))\Big) \right] \tag{5.22}$$

*for* $t \geq 0$ *and* $(i = 1, \cdots, 10)$, *where* $h_i = 0.5 * (11 - i), (i = 1, \cdots, 10)$, *and the delay* $\tau(t) = e^{-t} \in L^2$.

This network satisfies the conditions of Theorem 5.1 and Theorem 5.2. It is completely stable. Figure 5.2 shows the simulation results for the convergence of the trajectory with initial condition $\phi_i = 1(i = 1, \cdots, 10)$ for $t \in [-1, 0]$.

If suppose the delay $\tau(t) \equiv 0$ in the network (5.22), this network can perform as a "winner-take-all" function [15, 16, 62]. Figure 5.3 shows the corresponding delay-free simulation result to Figure 5.2.

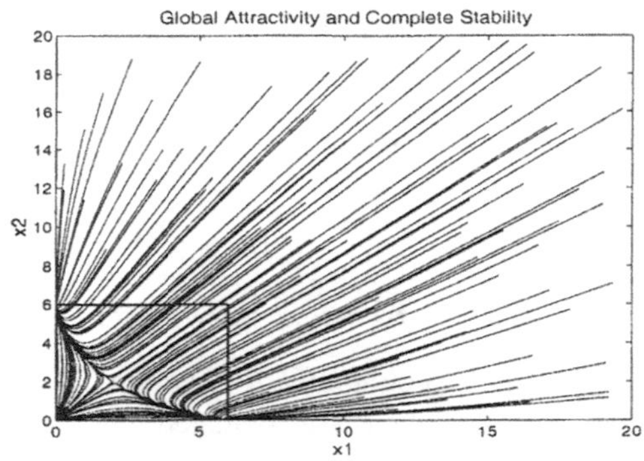


*Figure 5.1.* Global attractivity and complete convergence.

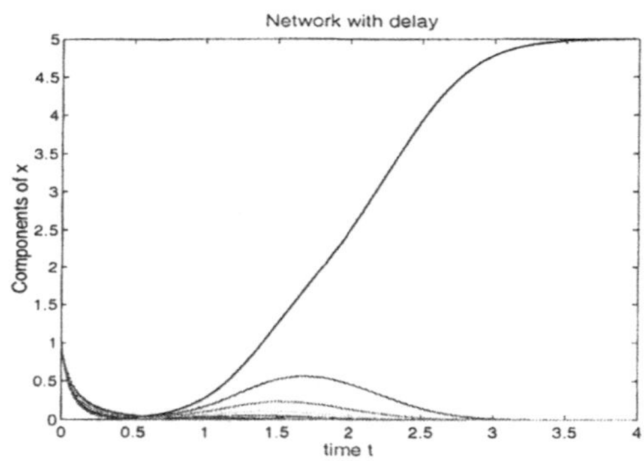


*Figure 5.2.* Convergence of network with delay $\tau(t) = e^{-t}$.

## 3. Monostability Analysis

This section focus on monostability of the network (5.1). Preliminaries will be given in subsection 3.1. In subsection 3.2, positive lower boundedness will be discussed. Subsection 3.3 will study exponential convergence. Subsection 3.4 will study asymptotic convergence. Simulation results will be given in subsection 3.5.

### 3.1 Preliminaries

In this section, some preliminaries for the network (5.1) which will be necessary in the analysis of the global convergence will be given.

DEFINITION 5.5 *The network (5.1) is said to be lower positive bounded, if for each trajectory $x(t) = (x_1(t), \cdots, x_n(t))^T$ of network (5.1), there exists a*

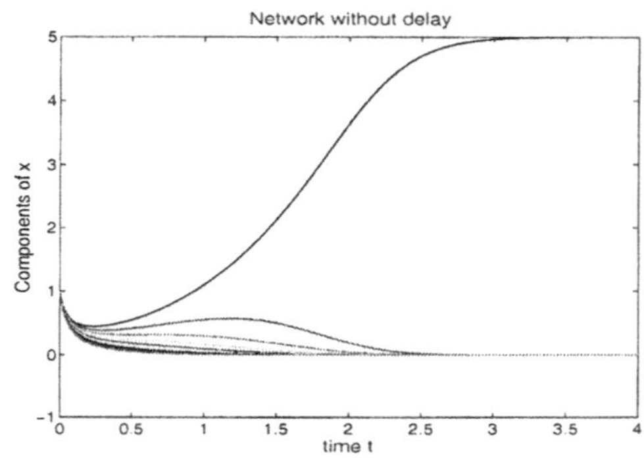


*Figure 5.3.* Convergence of network without delay.

*constant $c > 0$ such that*

$$x_i(t) \geq c, (i = 1, \cdots, n)$$

*for all $t \geq 0$.*

In the above definitions, the constant $c$ may depend on the trajectory $x(t)$. For different trajectories, the lower positive bounds may different. By a lower positive bound property, it implies that each component of a trajectory will never approach zero. This property has important applications in neural computation.

DEFINITION 5.6 *A vector $x^* = (x_1^*, \cdots, x_n^*)^T \in R^n$ is called a positive equilibrium of network (5.1), if $x_i^* > 0 (i = 1, \cdots, n)$ and*

$$h_i - x_i^* + \sum_{j=1}^{n} (a_{ij} + b_{ij}) x_j^* \equiv 0, (i = 1, \cdots, n). \quad (5.23)$$

Clearly, to calculate positive equilibrium points, it is sufficient to solve the linear algebra equation of (5.23). Since this section interested in monostability analysis of network (5.1) for positive equilibrium points, thus, throughout this section, the network (5.1) is assumed to have a unique positive equilibrium $x^* = (x_1^*, \cdots, x_n^*)^T \in R^n$.

Assume the initial condition as that of (5.2). Denote by $C^+$ the set of all the continuous functions which satisfy (5.2).

Denote

$$\| \phi - x^* \| = \max_{1 \leq i \leq n} \left[ \sup_{-\tau \leq s \leq 0} |\phi_i(s) - x_i^*| \right].$$

DEFINITION 5.7 *The network (5.1) is said to be globally exponentially stable, if for each trajectory $x(t)$ of (5.1), there exist constants $\epsilon > 0$ and $m \geq 1$, such that*

$$|x_i(t) - x_i^*| \leq m \cdot \| \phi - x^* \| \cdot e^{-\epsilon t}, (i = 1, \cdots, n)$$

*for $t \geq 0$.*

This definition of global exponential stability differs from the classical definition of global exponential stability. In our definition, the convergence rate $\epsilon$ may depend on each trajectory $x(t)$, whereas in the classical one, it is independent of the trajectories. Thus, the concept here is relatively weaker than the classical one, but it is interesting since it shows the exponential convergence of each trajectory.

DEFINITION 5.8 *The network (5.1) is called globally asymptotically stable, if for any $\epsilon > 0$, there exists a constant $\delta > 0$, such that $\| \phi - x^* \| < \delta$ implies that*

$$|x_i(t) - x_i^*| < \epsilon, (i = 1, \cdots, n)$$

*for all $t \geq 0$, and*

$$\lim_{t \to +\infty} |x_i(t) - x_i^*| = 0, (i = 1, \cdots, n).$$

LEMMA 5.2 *[221] Let $Q$ be any of a $n \times n$ matrix, then for any constant $k > 0$ and any symmetric positive matrix $S > 0$ that*

$$2x^T Q y \leq k x^T Q S^{-1} Q x + \frac{1}{k} y^T S y$$

*for all $x, y \in R^n$.*

## 3.2 Positive Lower Boundedness

Lemma 5.1 shows that each trajectory of (5.1), with initial condition in $C^+$, remains positive. However, as time approaches infinity, some components of a trajectory of (5.1) may approach zero. For an example, consider the two dimensional network

$$\begin{cases} \dot{x}_1(t) = x_1(t) \cdot [1 - x_1(t) - x_2(t)] \\ \dot{x}_2(t) = x_2(t) \cdot [2 - x_2(t) - x_1(t)] \end{cases} \tag{5.24}$$

for $t \geq 0$. By Theorem 5.1, the network (5.24) is bounded. Further, it is easy to derive from (5.24) that

$$x_1(t) = x_2(t) \cdot \frac{x_1(0)}{x_2(0)} \cdot e^{-t} \to 0$$

as $t \to +\infty$.

It is undesirable in some neural computations that some components of trajectories of (5.1) may approach zero. To solve this problem, in [15, 16, 62], a special $\epsilon$ term is introduced to the Lotka-Volterra network. [15, 16, 62] discussed the following Lotka-Volterra recurrent neural network

$$\tau \frac{dx_i(t)}{dt} = x_i \left( \gamma + W_i - x_i(t) - \lambda \sum_{j=1, j\neq i}^{n} x_j(t) \right) + \epsilon, (i = 1, \cdots, n) \tag{5.25}$$

for $t \geq 0$, where $\gamma$, $W_i$, $\lambda$ and $\epsilon > 0$ are constants. This model contains a $\epsilon$ term, its role is to prevent any $x_i(t)$ from becoming zero, so that losers and winners can interchange if the magnitudes of $\{W_i\}$ are changed occasionally [15]. However, introducing the $\epsilon$ term into the network (5.25) seems not reasonable since it is an artificial term not arising from the original membrane dynamics of competing neurons.

An interesting problem is to obtain the conditions such that each trajectory will never approach zero. Clearly, if the network has the property of lower positive bounded, then it is sufficient to guarantee that each trajectory has no component approaching zero. In this section, conditions for lower positive boundedness of the network (5.1) will be established.

THEOREM 5.3 *If there exists constants $\gamma_i > 0 (i = 1, \cdots, n)$ such that*

$$1 - a_{ii} - \frac{1}{\gamma_i} \sum_{j=1}^{n} \gamma_j \left[ |a_{ij}| \cdot (1 - \delta_{ij}) + |b_{ij}| \right] \geq 0, \quad (i = 1, \cdots, n),$$

*then, the network (5.1) is lower positive bounded.*

*Proof*: Given any $\phi \in C^+$, let $x(t)$ be the trajectory of (5.1) passing through $\phi$. By Lemma 5.1, $x_i(t) > 0, (i = 1, \cdots, n)$ for all $t \geq 0$.

Define

$$y_i(t) = \frac{|x_i(t) - x_i^*|}{\gamma_i}, (i = 1, \cdots, n),$$

for $t \geq -\tau$. Then, from (5.1),

$$\begin{aligned} D^+ y_i(t) \leq \;& x_i(t) \Bigg[ (-1 + a_{ii}) y_i(t) + \frac{1}{\gamma_i} \sum_{j=1}^{n} \gamma_j \Big( |a_{ij}| \cdot (1 - \delta_{ij})\, y_j(t) \\ & + |b_{ij}| y_j(t - \tau_{ij}(t)) \Big) \Bigg], \end{aligned} \tag{5.26}$$

for $t \geq 0$.

Since the condition of Theorem 5.3 implies the condition of Theorem 5.1, $x_i(t) (i = 1, \cdots, n)$ and then $y_i(t) (i = 1, \cdots, n)$ must be bounded. Define

$$\lim_{t \to +\infty} \sup y_i(t) = \sigma_i, (i = 1, \cdots, n).$$

Clearly, $0 \leq \sigma_i < +\infty (i = 1, \cdots, n)$. Suppose

$$\sigma_l = \max_{1 \leq i \leq n} \{\sigma_i\}.$$

It will prove that $\sigma_l = 0$.

Denote

$$\beta_i \triangleq 1 - a_{ii} - \frac{1}{\gamma_i} \sum_{j=1}^{n} \gamma_j \left[ |a_{ij}|(1 - \delta_{ij}) + |b_{ij}| \right], (i = 1, \cdots, n).$$

Clearly, $1 - a_{ii} \geq \beta_i > 0 (i = 1, \cdots, n)$.

If $\sigma_l \neq 0$, i.e., $\sigma_l > 0$, then a small constant $\epsilon > 0$ can be chosen such that

$$0 < \epsilon \leq \frac{\beta_l \sigma_l}{4(1 - a_{ll}) - 2\beta_l}$$

By the basic property of upper limit, there exists a $t_1 \geq 0$ such that

$$y_i(t - \tau_{ij}(t)) \leq \sigma_i + \epsilon \leq \sigma_l + \epsilon, (i, j = 1, \cdots, n) \tag{5.27}$$

for all $t \geq t_1$.

Next, it proves that there exists a $t_2 \geq t_1$ such that

$$D^+ y_l(t) < 0 \tag{5.28}$$

for all $t \geq t_2$. If this is not true, then for any $t_2 \geq t_1$ there must exist a $t_3 > t_2$ such that

$$D^+ y_l(t_3) \geq 0 \tag{5.29}$$

and

$$y_l(t_3) \geq \sigma - \epsilon. \tag{5.30}$$

From (5.26), (5.27) and (5.30), it follows that

$$\begin{aligned} D^+ y_l(t_3) &\leq x_l(t_3) \Bigg[ -(1 - a_{ll})(\sigma_l - \epsilon) \\ &\quad + \frac{1}{\gamma_l} \sum_{j=1}^{n} \gamma_j \left( |a_{lj}|(1 - \delta_{lj}) + |b_{lj}| \right) (\sigma_l + \epsilon) \Bigg] \\ &= x_l(t_3) \left[ -\beta_l \sigma_l + (2(1 - a_{ll}) - \beta_l) \epsilon \right] \\ &< -\frac{\beta_l \sigma_l}{2} \cdot x_l(t_3) \\ &< 0 \end{aligned}$$

which contradicts (5.29). Thus, (5.28) is true. By (5.28), $y(t)$ is monotone decreasing when $t$ is sufficiently large, thus the limit of $y(t)$ must exist, i.e.,

$$\lim_{t\to+\infty} y_l(t) = \lim_{t\to+\infty} \sup y_l(t) = \sigma_l.$$

By the definition of limit, there exists a $t_4 \geq t_2$ such that

$$\sigma_l - \epsilon \leq y_l(t - \tau_{lj}(t)) \leq \sigma_l + \epsilon, (j = 1, \cdots, n) \tag{5.31}$$

for $t \geq t_4$.

From (5.1) and (5.31),

$$\begin{aligned} D^+ \left| \ln \frac{x_l(t)}{x_l^*} \right| &\leq \gamma_l \Bigg[ -(1 - a_{ll}) y_l(t) \\ &\quad + \frac{1}{\gamma_l} \sum_{j=1}^{n} \gamma_j \left( |a_{lj}|(1 - \delta_{lj}) y_j(t) + |b_{lj}| y_j(t - \tau_{lj}(t)) \right) \Bigg] \\ &\leq \gamma_l \Bigg[ -(1 - a_{ll})(\sigma_l - \epsilon) \\ &\quad + \frac{1}{\gamma_l} \sum_{j=1}^{n} \gamma_j \left( |a_{lj}|(1 - \delta_{lj}) + |b_{lj}| \right) (\sigma_l + \epsilon) \Bigg] \\ &= -\frac{\beta_l \sigma_l \gamma_l}{2} \end{aligned}$$

for all $t \geq t_4$. Then, it follows that

$$\begin{aligned} \left| \ln \frac{x_l(t)}{x_l^*} \right| &\leq \left| \ln \frac{x_l(t_4)}{x_l^*} \right| - \frac{\beta_l \sigma_l \gamma_l}{2} \cdot (t - t_4) \\ &\to -\infty \end{aligned}$$

as $t \to +\infty$. This is a contradiction, then it must be $\sigma_l = 0$. Thus, it follows that

$$\lim_{t\to+\infty} x_i(t) = x_i^*, (i = 1, \cdots, n).$$

Since $x_i^* > 0 (i = 1, \cdots, n)$, clearly, there must exist a $t_5 \geq 0$ such that

$$x_i(t) \geq x_i^*/2, (i = 1, \cdots, n).$$

for $t \geq t_5$. Then,

$$x_i(t) \geq \min_{1\leq j\leq n} \left\{ \frac{x_j^*}{2}, \inf_{0\leq s\leq t_5} \{x_j(s)\} \right\} > 0, (i = 1, \cdots, n)$$

for all $t \geq 0$. That is, the network (5.1) is lower positive bounded. The proof is completed.

Theorem 5.3 shows that under suitable conditions, each component of a trajectory will never become zero. Thus, the $\epsilon$ term in the network model (5.25) can be removed under some conditions.

## 3.3 Exponential Convergence

In this section, conditions for global exponential convergence of the network (5.1) will be established. The network (5.1) is a nonlinear system. Global Lipshitz linearization is a well known method for establishing exponential convergence of nonlinear systems. However, since the network (5.1) contains product terms in the equation, it is difficult to apply global Lipshitz linearization. Moreover, the methods of [220] and [224] for analyzing exponential convergence of Hopfield neural networks are also not applicable to the network (5.1). In this section, with the property of lower positive boundedness, it is able to establish exponential convergence conditions for the network (5.1).

THEOREM 5.4 *If there exist constants* $\gamma_i > 0 (i = 1, \cdots, n)$ *such that*

$$1 - a_{ii} - \frac{1}{\gamma_i}\sum_{j=1}^{n}\gamma_j\left[|a_{ij}| \cdot (1 - \delta_{ij}) + |b_{ij}|\right] > 0, \quad (i = 1, \cdots, n),$$

*then, the network (5.1) is globally exponentially stable.*

*Proof*: Given any $\phi \in C^+$. Let $x(t)$ be the trajectory of (5.1) passing through $\phi$. By Theorem 5.3, the network (5.1) is lower positive bounded, then there exists a constant $c > 0$ such that

$$x_i(t) \geq c, (i = 1, \cdots, n)$$

for $t \geq 0$.

Define functions

$$f_i(s) = -s + c\left[1 - a_{ii} - \frac{1}{\gamma_i}\sum_{j=1}^{n}\gamma_j\left(|a_{ij}|(1 - \delta_{ij}) + e^{s\tau}|b_{ij}|\right)\right]$$

for $s \in (-\infty, +\infty)$ and $(i = 1, \cdots, n)$. Clearly, $f_i(s)(i = 1, \cdots, n)$ are continuous functions with the property that $f_i(0) > 0 (i = 1, \cdots, n)$. Then, by continuity, there must exist a small constant $\epsilon > 0$ such that

$$f_i(\epsilon) = -\epsilon + c\left[1 - a_{ii} - \frac{1}{\gamma_i}\sum_{j=1}^{n}\gamma_j\left(|a_{ij}|(1 - \delta_{ij}) + e^{\epsilon\tau}|b_{ij}|\right)\right] > 0 \quad (5.32)$$

for $(i = 1, \cdots, n)$.

Define

$$z_i(t) = \frac{e^{\epsilon t}|x_i(t) - x_i^*|}{\gamma_i}, (i = 1, \cdots, n),$$

for $t \geq -\tau$. Then, from (5.1),

$$\begin{aligned} D^+ z_i(t) \leq\ & \epsilon z_i(t) + x_i(t)\Bigg[(-1 + a_{ii})z_i(t) \\ & + \frac{1}{\gamma_i}\sum_{j=1}^{n} \gamma_j\Big(|a_{ij}| \cdot (1 - \delta_{ij})\, z_j(t) \qquad (5.33) \\ & + e^{\epsilon\tau}|b_{ij}|z_j(t - \tau_{ij}(t))\Big)\Bigg], \qquad (5.34) \end{aligned}$$

for $t \geq 0$.

It will prove that

$$z_i(t) < 2 \cdot \max_{1 \leq i \leq n} \left\{\frac{1}{\gamma_i}\right\} \cdot \| \phi - x^* \| \triangleq \Pi, (i = 1, \cdots, n), \qquad (5.35)$$

for all $t \geq 0$. Clearly, $z_i(t) < \Pi (i = 1, \cdots, n)$ for $t \in [-\tau, 0]$. If (5.35) is not true, then, there must exist a $t_1 > 0$ such that $D^+ z_i(t_1) \geq 0$ and

$$z_i(t_1) = \Pi; \quad z_j(t) \begin{cases} < \Pi, & -\tau \leq t < t_1, \quad i = j \\ \leq \Pi, & -\tau \leq t \leq t_1, \quad i \neq j \end{cases}. \qquad (5.36)$$

However, from (5.34), (5.36) and (5.32), it follows that

$$\begin{aligned} D^+ z_i(t_1) \leq\ & \epsilon\Pi + x_i(t_1)\Bigg( -1 + a_{ij} \\ & + \frac{1}{\gamma_i}\sum_{j=1}^{n} \gamma_j\Big(|a_{ij}| \cdot (1 - \delta_{ij}) + e^{\epsilon\tau}|b_{ij}|\Big)\Bigg)\Pi \\ \leq\ & \Pi\Bigg[\epsilon + c\Bigg( -1 + a_{ij} \\ & + \frac{1}{\gamma_i}\sum_{j=1}^{n} \gamma_j\Big(|a_{ij}| \cdot (1 - \delta_{ij}) + e^{\epsilon\tau}|b_{ij}|\Big)\Bigg)\Bigg] \\ =\ & -f_i(\epsilon) \cdot \Pi \\ <\ & 0. \end{aligned}$$

This poses a contradiction and it implies that (5.35) is true. By (5.35), it gives that

$$|x_i(t) - x_i^*| \leq m \cdot \| \phi - x^* \| \cdot e^{-\epsilon t}, (i = 1, \cdots, n)$$

for $t \geq 0$, where

$$m = 2 \cdot \max_{1 \leq j \leq n} \{\gamma_j\} \cdot \max_{1 \leq j \leq n} \left\{ \frac{1}{\gamma_j} \right\} > 1.$$

This shows the network (5.1) is globally exponentially stable and the proof is completed.

Taking $\gamma_i = 1 (i = 1, \cdots, n)$ in Theorem 5.4, then it gives especially simple results.

COROLLARY 5.3 *If*

$$1 - a_{ii} - \sum_{j=1}^{n} \left( |a_{ij}| \cdot (1 - \delta_{ij}) + |b_{ij}| \right) > 0, \quad (i = 1, \cdots, n)$$

*then the network (5.1) is globally exponentially stable.*

Theorem 5.4 gives a condition for global exponential convergence of network (5.1). The condition is clearly independent of the delays of the networks. It is quite easy to check. The exponential convergence rate of the network (5.1) can be estimated from (5.32).

## 3.4 Asymptotic Convergence

In the last subsection, several sufficient conditions for network (5.1) to be exponentially convergent have been obtained. However, since these conditions are only sufficient but not necessary, there may exist certain networks of the model (5.1) which are convergent but do not satisfy the conditions of Theorem 5.4. In this subsection, we will construct novel Lyapunov functions and functionals to establish different conditions for asymptotic convergence of the network (5.1). Since the network (5.1) contains nonlinear product terms and delays, the Lyapunov function and functional should have the ability to overcome the problems caused by the product terms and delays.

Suppose, in this subsection, that for each $i (i = 1, \cdots, n)$, the delays $\tau_{ij}(t) (j = 1, \cdots, n)$ independent of $j$, i.e. $\tau_{i1}(t) = \cdots = \tau_{in}(t)$, and denote

$$\tau_i(t) \triangleq \tau_{ij}(t), (i, j = 1, \cdots, n)$$

for $t \geq 0$. For notation convenience, denote

$$x^T(t - \tau(t)) = (x_1(t - \tau_1(t)), \cdots, x_n(t - \tau_n(t)))^T.$$

THEOREM 5.5 *Suppose the delays $\tau_j(t) (j = 1, \cdots, n)$ are differentiable, and*

$$\sigma \triangleq \max_{1 \leq i \leq n} \left\{ \sup_{t \geq 0} \{\dot{\tau}_i(t)\} \right\} < 1.$$

*If there exists a diagonal matrix $D > 0$ and a constant $k > 0$ such that*

$$-D + \frac{DA + A^T D + kDBB^T D}{2} + \frac{1}{2k(1-\sigma)} I < 0$$

*then, the network (5.1) is globally asymptotically stable.*

*Proof*: Given any $\phi \in C^+$, let $x(t) = (x_1(t), \cdots, x_n(t))^T$ be the trajectory of (5.1) passing through $\phi$. By, Lemma 5.1, $x_i(t) > 0 (i = 1, \cdots, n)$ for all $t \geq 0$.

Suppose $D = diag(d_1, \cdots, d_n)$, then $d_i > 0 (i = 1, \cdots, n)$. Constructing a Lyapunov function

$$V_1(t) = \sum_{i=1}^{n} d_i \int_{x_i^*}^{x_i(t)} \frac{\theta - x_i^*}{\theta} d\theta \tag{5.37}$$

for $t \geq 0$. Clearly, $V_1(t) \geq 0$ for all $t \geq 0$.

It follows from (5.37), (5.1) and Lemma 5.2 that

$$\begin{aligned}
\dot{V}_1(t) &= \sum_{i=1}^{n} d_i [x_i(t) - x_i^*] \cdot \frac{\dot{x}_i(t)}{x_i(t)} \\
&= [x(t) - x^*]^T D \Big[(-I + A)[x(t) - x^*] + B[x(t - \tau(t)) - x^*]\Big] \\
&= [x(t) - x^*]^T \left[ -D + \frac{DA + A^T D + kDBB^T D}{2} \right] [x(t) - x^*] \\
&\quad + \frac{1}{2k} \parallel x(t - \tau(t)) - x^* \parallel^2 \\
&\leq -\left( \lambda + \frac{1}{2k(1-\sigma)} \right) \parallel x(t) - x^* \parallel^2 \\
&\quad + \frac{1}{2k} \parallel x(t - \tau(t)) - x^* \parallel^2
\end{aligned} \tag{5.38}$$

for $t \geq 0$, where $\lambda$ is the minimum eigenvalue of the matrix

$$D - \frac{DA + A^T D - kDBB^T D}{2} - \frac{1}{2k(1-\sigma)} I.$$

Clearly, $\lambda > 0$.

Define a Lyapunov functional

$$V(t) = V_1(t) + \frac{1}{2k(1-\sigma)} \sum_{i=1}^{n} \int_{t-\tau_i(t)}^{t} [x_i(s) - x_i^*]^2 \, ds \tag{5.39}$$

for $t \geq 0$.

Then, from (5.39) and (5.38), it follows that

$$\begin{aligned}\dot{V}(t) &= \dot{V}_1(t) + \frac{1}{2k(1-\sigma)} \| x(t) - x^* \|^2 \\ &\quad -\frac{1}{2k} \| x(t-\tau(t)) - x^* \|^2 \\ &\leq -\lambda \| x(t) - x^* \|^2 \end{aligned} \tag{5.40}$$

for all $t \geq 0$. Then,

$$d_i \int_{x_i^*}^{x_i(t)} \frac{\theta - x_i^*}{\theta} d\theta \leq V_1(t) \leq V(t) \leq V(0) \leq \Pi \cdot \| \phi - x^* \|^2 \tag{5.41}$$

for $t \geq 0$ and $(i = 1, \cdots, n)$, where

$$\Pi = \frac{1}{\lambda} \cdot \left[ \sum_{i=1}^{n} \frac{d_i}{2x_i^*} + \frac{n\tau}{2k(1-\sigma)} \right].$$

Clearly, for any $\epsilon > 0$, there exists a $\delta > 0$ such that $\| \phi - x^* \| \leq \delta$ implies that

$$|x_i(t) - x_i^*| \leq \epsilon, (i = 1, \cdots, n)$$

for $t \geq 0$.

From (5.40), it gives

$$\int_0^t \| x(s) - x^* \|^2 ds \leq \frac{V(0) - V(t)}{\lambda}$$

for $t \geq 0$, and so

$$\int_0^{+\infty} \| x(s) - x^* \|^2 ds < +\infty. \tag{5.42}$$

By (5.41), $x(t)$ is bounded. Using (5.1), then $\dot{x}(t)$ is bounded. Thus, $x(t)$ is uniformly continuous on $[0, +\infty)$, and then it follows from (5.42) that

$$\lim_{t \to +\infty} \| x(t) - x^* \| = 0.$$

This shows the network (5.1) is globally asymptotically stable. The proof is completed.

The Lyapunov function which we have constructed in (5.37) can efficiently overcome the problem of nonlinear product terms in (5.1). The role of Lyapunov functional in (5.39) is to remove the delays.

Conditions of Theorem 5.5 are different from that of Theorem 5.4. Consider the following example,

$$\dot{x}(t) = x(t) \cdot \left[ -3x(t) + 2x\left(t - \frac{9\sin^2 t}{10}\right) \right]$$

for $t \geq 0$. It satisfies the conditions of Theorem 5.4 but does not fulfill conditions of Theorem 5.5. In next section, we will give an example which satisfies conditions of Theorem 5.5 but cannot be checked by Theorem 5.4. Thus, conditions between Theorem 5.4 and Theorem 5.5 do not pose implicative relationships in general. They can be used in different situations.

If the delays are constants in the network (5.1), Theorem 5.4 gives a simple results.

COROLLARY 5.4 *Suppose the delays $\tau_j(t)(j = 1, \cdots, n)$ are constants. If there exists a diagonal matrix $D > 0$ and a constant $k > 0$ such that*

$$-D + \frac{DA + A^T D + kDBB^T D}{2} + \frac{1}{2k} I < 0$$

*then, the network (5.1) is globally asymptotically stable.*

## 3.5 Simulation Results

In this section, some examples will be employed to further illustrate the above theory. For simulation convenience, suppose that the networks contain only one constant delay.

EXAMPLE 5.3 *Consider the following ten-dimensional Lotka-Volterra neural network with one delay,*

$$\begin{cases} \dot{x}_1(t) = x_1(t)\left[h_1 - x_1(t) - \dfrac{1}{10} \cdot x_2(t-1)\right] \\ \dot{x}_i(t) = x_i(t)\left[h_i - x_i(t) - \dfrac{1}{10} \cdot \left(x_{i-1}(t) + x_{i+1}(t-1)\right)\right], (i = 2, \cdots, 9) \\ \dot{x}_{10}(t) = x_{10}(t)\left[h_{10} - x_{10}(t) - \dfrac{1}{10} \cdot x_9(t)\right], \end{cases} \tag{5.43}$$

*for $t \geq 0$, where $h_i = 11 - i, (i = 1, \cdots, 10)$, and the delay $\tau = 1$ is a constant.*

This network satisfies the conditions of Theorem 5.4. It is globally exponentially stable. Figure 5.4 shows the simulation results for the convergence of the trajectory with initial condition $\phi$ as

$$(0.9204, 2.7718, 2.7091, 2.2663, 1.2680, 3.8782, 3.9961, 4.1375, 2.8919, 3.2742)^T$$

for $t \in [-1, 0]$.

The network converges to $x^*$ as

$$(8.7827, 6.0863, 5.7857, 4.9851, 4.2888, 3.5708, 2.8573, 2.1428, 1.4286, 0.7143)^T.$$

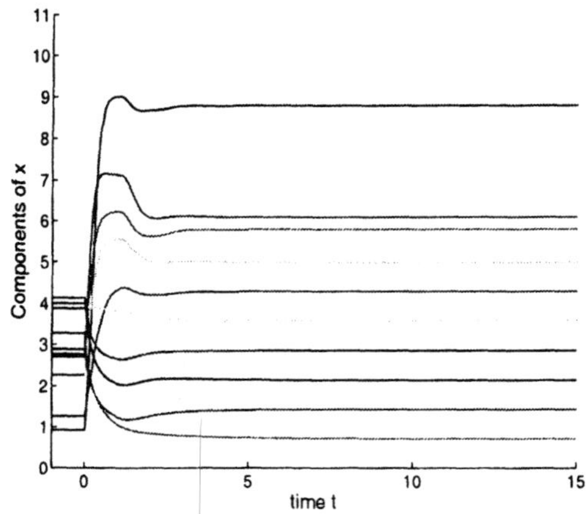


*Figure 5.4.* Exponential convergence of network (5.43) with delay $\tau = 1$. The trajectory starts from a randomly selected constant vector $\phi$ which is clearly in $C^+$. It can be observed that the network converges quickly to a fixed point.

EXAMPLE 5.4 *Consider the two-dimensional network*

$$\begin{cases} \dot{x}_1(t) = x_1(t)\Big[6 - x_1(t) + \Big( -2x_1(t) + x_2(t) - x_1(t-1) + x_2(t-1)\Big)\Big] \\ \dot{x}_2(t) = x_2(t)\Big[3 - x_2(t) + \Big( -x_1(t) - x_2(t) + x_1(t-1) + x_2(t-1)\Big)\Big] \end{cases} \tag{5.44}$$

*for* $t \geq 0$. *The delay* $\tau = 1$.

This network does not satisfy the conditions of Theorem 5.4. Next, invoke Theorem 5.5 instead. Clearly,

$$A = \begin{bmatrix} -2 & 1 \\ -1 & -1 \end{bmatrix}, \quad B = \begin{bmatrix} -1 & 1 \\ 1 & 1 \end{bmatrix}.$$

Choose $D = I$ and $k = 1/2$, then

$$-D + \frac{DA + A^T D + kDBB^T D}{2} + \frac{1}{2k} I = \begin{bmatrix} -1.5 & 0 \\ 0 & -0.5 \end{bmatrix} < 0.$$

By Theorem 5.5, the network (5.44) is globally asymptotically stable.

Figure 5.5 shows the convergence of the network with 20 trajectories starting from randomly selected initial constant vectors. It can be observed that each trajectory converges to the positive equilibrium $(3, 3)$.

Conditions given in Theorem 5.5 depend on network's delays if the the delays are not constants. Theorem 5.5 is interesting because it provides convergence

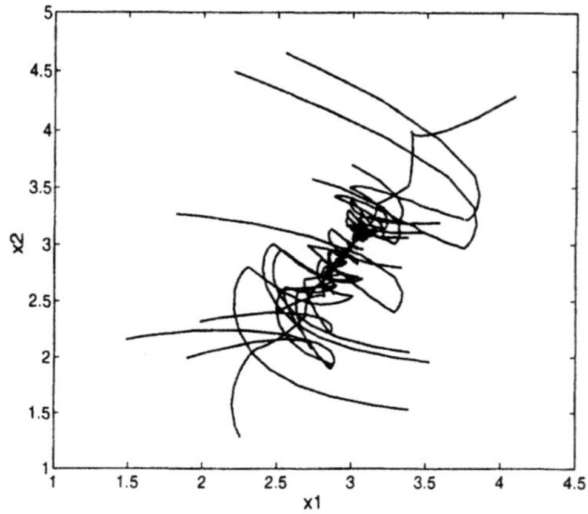


*Figure 5.5.* Convergence of network (5.44) with 20 trajectories. Each trajectory starts from a randomly selected constant vector.

conditions different from that of Theorem 5.4. However, since the conditions of Theorem 5.5 depend on delays, they are relatively conservative. Our conjecture is that in Theorem 5.5, the conditions could be replaced by

$$-D + \frac{DA + A^T D + kDBB^T D}{2} + \frac{1}{2k} I < 0$$

and the result still holds. This would be a good delays independent condition. Unfortunately, this point cannot be proven at present. It would be an interesting problem to be studied in future.

## 4. Conclusions

In this chapter, convergence of general recurrent Lotka-Volterra RNNs with variable delays is studied. Both multistability and monostability are discussed. LV RNNs have many clear dynamical properties and the some hardware implementations are now available [16]. It is believed that LV RNNs will have more applications in future.

Chapter 6

# DELAYED RECURRENT NEURAL NETWORKS WITH GLOBAL LIPSCHITZ ACTIVATION FUNCTIONS

## 1. Introduction

It is well known that activation functions of RNNs are important characters of RNNs. Usually, activation functions are nonlinear functions, thus they make RNNs be described by nonlinear systems. Since they represent some structural parts of the the nonlinear systems, they crucially decide the dynamical properties of RNNs.

As known, delays are important parameters of RNNs. They also affect the dynamical properties of RNNs. A stable RNN may become unstable by adding some delays. Thus, it is important to study the dynamical properties of RNNs with delays. Either in biological or artificial neural networks, state time delays exist in the networks. RNNs without delays are ideal cases but not realistically in practice.

This chapter studies the convergence of several types of continuous time delayed RNNs with activation functions satisfying global Lipschitz linearization. It has been long known that Lipschitz linearization is an effective method for analyzing convergence of nonlinear systems.

### 1.1 Global Lipschitz Activation Functions

Let $g : R \to R$ be an activation function. $g$ is called a global Lipschitz activation function (GLAF), if it is a continuous, monotonically increasing, and nonconstant function, and it satisfies the following global Lipschitz linearization

$$0 \leq \frac{g(\xi) - g(\eta)}{\xi - \eta} \leq l, \tag{6.1}$$

for all $\xi, \eta \in R$ and $\xi \neq \eta$, where $l > 0$ is a constant. Denote by $S_{GLAF}$ the set of all GLAG functions.

Clearly, a GLAF may be unbounded. Many frequently used activation functions belong to $S_{GLAF}$, for examples, $1/(1+e^{-\theta})$, $(2/\pi)arctan(\theta)$, $max(0,\theta)$ and $(|\theta+1|-|\theta-1|)/2$, where $\theta \in R$, are all GLAFs.

It is easy to show that (6.1) equivalent to

$$0 \le D^+g(s) \le l \tag{6.2}$$

for all $s \in R$.

LEMMA 6.1 *If $g \in S_{GLAF}$, then*

$$[g(\theta)-g(\eta)]^2 \le l \cdot (\theta-\eta)\,[g(\theta)-g(\eta)] \le l^2\,(\theta-\eta)^2$$

*for all $\theta, \eta \in R$.*

LEMMA 6.2 *If $g \in S_{GLAF}$, then*

$$\int_\eta^\xi [g(\theta)-g(\eta)]\,d\theta \le (\xi-\eta)\,[g(\xi)-g(\eta)]$$

*for all $\xi, \eta \in R$.*

The proof of Lemma 6.1 and Lemma 6.2 are directly from the definition of GLAF.

LEMMA 6.3 *If $g \in S_{GLAF}$, then*

$$\int_\eta^\xi [g(\theta)-g(\eta)]\,d\theta \ge \frac{1}{2l}\,[g(\xi)-g(\eta)]^2$$

*for all $\xi, \eta \in R$.*

*Proof*: Define a continuous function

$$W(\xi) = \int_\eta^\xi [g(s)-g(\eta)]\,ds - \frac{1}{2l}\,[g(\xi)-g(\eta)]^2.$$

Then, it follows that

$$D^+W(\xi) = [g(\xi)-g(\eta)]\left[1-\frac{D^+g(\xi)}{l}\right]^2.$$

By (6.2), it follows that

$$D^+W(\xi)\begin{cases} \ge 0, & \text{if } \xi > \eta \\ \le 0, & \text{if } \xi < \eta \\ = 0, & \text{if } \xi = \eta. \end{cases}$$

This shows that $\xi = \eta$ is the minimum point of the function $E(\xi)$. Then, for any $\xi \in R$, it follows that $E(\xi) \ge E(\eta) = 0$. Thus,

$$\int_\eta^\xi [g(s)-g(\eta)]\,ds \ge \frac{1}{2l}\,[g(\xi)-g(\eta)]^2,$$

for all $\xi, \eta \in R$. This completes the proof.

## 1.2 Functional Differential Equations

Since the models of RNNs with delays belong to functional differential equations in mathematics, this subsection gives a brief introduction to functional differential equations. More details about functional differential equations can be found in [78].

Suppose that $\tau \geq 0$ is a constant. Let $\mathcal{C} = \mathcal{C}([-\tau, 0], R^n)$ be the Banach space of continuous functions mapping the interval $[-\tau, 0]$ into $R^n$ with the topology of uniform convergence. For any $\phi \in \mathcal{C}$, define a normal

$$\|\phi\| = \sup_{-\tau \leq \theta \leq 0} |\phi(\theta)|$$

where $|\cdot|$ is an Eclidian normal in $R^n$. It should not be confused that $\|\cdot\|$ and $|\cdot|$ are different normal which are defined on different space.

Let $f : R \times \mathcal{C} \to R^n$ be a continuous functional. The functional differential equation is defined by

$$\dot{x}(t) = f(t, x_t), \tag{6.3}$$

for $t \geq 0$, where $\cdot$ represents the right-hand derivative. For any $\phi \in \mathcal{C}$, let $x(t, \phi)$ denote the trajectory of (6.3) starting from $\phi$.

A point $x^* \in R^n$ (clearly, $x^* \in \mathcal{C}$) is called an equilibrium of (6.3), if $f(t, x^*) \equiv 0$ for all $t \geq 0$.

DEFINITION 6.1 *An equilibrium $x^*$ of (6.3) is called stable, if given any $\epsilon > 0$, there exists a constant $\delta > 0$ such that $\|\phi - x^*\| < \delta$ implies that $|x(t, \phi) - x^*| < 0$ for all $t \geq 0$.*

DEFINITION 6.2 *The equation (6.3) is called globally asymptotically stable (GAS), if (6.3) has a unique equilibrium $x^*$ which is stable, and given any $\phi \in \mathcal{C}$, it holds that*

$$x(t, \phi) \to x^*$$

*as $t \to +\infty$.*

DEFINITION 6.3 *The equation (6.3) is called globally exponentially stable (GEA), if (6.3) has a unique equilibrium $x^*$, and there exist constants $\epsilon > 0$ and $\pi \geq 1$ such that*

$$|x(t, \phi) - x^*| \leq \pi \| \phi - x^* \| e^{-\epsilon t}$$

*for all $t \geq 0$ and any $\phi \in \mathcal{C}$.*

## 1.3 Organisations

The rest of this chapter is organised as follows. In Section 2, global exponential stability of RNNs with constant delays are studied. It will establish conditions for GES and estimation of GES convergence rate. In Section 3, GES of RNNs with variable delays will be studied. A special model of RNNs with delays will be discussed in Section 4. Absolute stability and absolute periodicity of RNNs with delays will be studied in Section 5. In Section 6, bidirectional associative memory RNNs with delays will be analyzed.

## 2. RNNs with Constant Delays

### 2.1 Preliminaries

The model of RNNs with constant delays is described by

$$c_i \frac{du_i(t)}{dt} = -d_i u_i(t) + \sum_{j=1}^{n} \Big[a_{ij} g_j(u_j(t)) + b_{ij} g_j(u_j(t-\tau_j))\Big] + I_i \tag{6.4}$$

where $c_i > 0$, $d_i > 0$, $a_{ij}$, $b_{ij}$ and $I_i$ are constants, $\tau_j \geq 0 (j = 1, \cdots, n)$ represent delays. Denote $\tau = \max_{1 \leq j \leq n}(\tau_j)$. The activation functions $g_i \in G_{GLAF}(i = 1, \cdots, n)$, i.e, there exist constants $l_i > 0 (i = 1, \cdots, n)$ such that

$$0 \leq \frac{g_i(\theta) - g_i(\eta)}{\theta - \eta} \leq l_i$$

for all $\theta, \eta \in R$ and $\theta \neq \eta$. Denote $L = diag\,(l_1, \cdots, l_n)_{n \times n}$.

Denote

$$\begin{cases} C = diag(c_1, \cdots, c_n) \\ D = diag(d_1, \cdots, d_n) \\ A = (a_{ij})_{n \times n} \\ B = (b_{ij})_{n \times n} \\ I = (I_1, \cdots, I_n)^T \\ u(t) = (u_1(t), \cdots, u_n(t))^T \\ G(u(t)) = (g_1(u_1(t)), \cdots, g_n(u_n(t)))^T \\ G(u(t-\tau)) = (g_1(u_1(t-\tau_1)), \cdots, g_n(u_n(t-\tau_n)))^T \end{cases}$$

then, (6.4) can be written in a compact vector form

$$C \frac{du(t)}{dt} = -Du(t) + AG(u(t)) + BG(u(t-\tau)) + I.$$

A vector $u^* = (u_1^*, \cdots, u_n^*)^T \in R^n$ is called an equilibrium of the network (6.4), if it satisfies

$$-Du^* + (A + B)\, G(u^*) + I = 0.$$

Obviously, the existence of equilibrium points depends on the activation functions $g_i (i = 1, \cdots, n)$. If each $g_i$ is bounded, it is easy to show that the network has at least one equilibrium. However, if the bounded condition of the activation functions is removed, the network may not have any equilibrium [57]. The following lemma gives a condition to guarantee the existence of an equilibrium of the network (6.4) without the bounded condition of activation functions. It was first proved in [60] with an additional condition that $g_i(0) = 0 (i = 1, \cdots, n)$. Later in [115], it was showed that this assumption is unnecessary and then it can be removed.

LEMMA 6.4 *[115] If*

$$-DL^{-1} + \frac{A + A^T + B + B^T}{2} < 0,$$

*the network (6.4) has a unique equilibrium.*

Using Lemma 6.4, the next result can be derived. Through this section, $E$ will be used to denote the identity matrix.

LEMMA 6.5 *If there exist constants $\sigma > 0$ and $k > 0$ such that*

$$-(D - \sigma C)L^{-1} + \frac{A + A^T}{2} + kE + \frac{e^{\sigma\tau}}{4k}BB^T < 0,$$

*the network (6.4) has a unique equilibrium.*

*Proof*: Given any $x \in R^n$, it follows that

$$\begin{aligned} & x^T \left[\sigma C L^{-1} + kE + \frac{e^{\sigma\tau}}{4k}BB^T - \frac{B + B^T}{2}\right] x \\ \geq\ & x^T \left[kE + \frac{1}{4k}BB^T - \frac{B + B^T}{2}\right] x \\ =\ & \frac{1}{4k}\left[B^T x - 2kx\right]^T \left[B^T x - 2kx\right] \\ \geq\ & 0 \end{aligned}$$

then,

$$\begin{aligned} & x^T \left[-DL^{-1} + \frac{A + A^T + B + B^T}{2}\right] x \\ \leq\ & x^T \left[-(D - \sigma C)L^{-1} + \frac{A + A^T}{2} + kE + \frac{e^{\sigma\tau}}{4k}BB^T\right] x. \end{aligned}$$

This shows the matrix

$$-DL^{-1} + \frac{A + A^T + B + B^T}{2}$$

is negative definite. The result now follows from Lemma 6.4. The proof is completed.

The initial condition of the network (6.4) is assumed to be

$$u_i(t) = \phi_i(t), -\tau \le t \le 0, i = 1, \cdots, n$$

where $\phi = (\phi_1, \cdots, \phi_n)^T \in C$. Let $u^* = (u_1^*, \cdots, u_n^*)^T$ be an equilibrium of (6.4), Denote

$$\| \phi - u^* \| = \sup_{-\tau \le \theta \le 0} \sqrt{\sum_{i=1}^{n} |\phi_i(\theta) - u_i^*|^2}.$$

DEFINITION 6.4 *The network (6.4) is said to be globally exponentially stable, if the network (6.4) has a unique equilibrium $(u_1^*, \cdots, u_n^*)$, and there exist constants $\epsilon > 0$ and $\pi \ge 1$ such that*

$$|u_i(t) - u_i^*| \le \pi \| \phi - u^* \| e^{-\epsilon t}, (i = 1, \cdots, n)$$

*for all $t \ge 0$ and $\phi \in C$.*

## 2.2 Convergence Rate Estimate

In this subsection, the estimate of global exponential convergence rate of network (6.4) will be discussed. In some applications, it is important to control the convergence speed of neural networks. Knowing the convergence rate provides a way to solve this problem. On the other hand, convergence rate analysis gives intrinsic insight to understand the dynamic behaviours of neural networks.

THEOREM 6.1 *If there exist two positive constants $\sigma$ and $k$ such that $0 < \sigma \le d_i/c_i (i = 1, \cdots, n)$ and*

$$-(D - \sigma C)L^{-1} + \frac{A + A^T}{2} + kE + \frac{e^{\sigma\tau}}{4k} BB^T < 0,$$

*the network (6.4) has a unique equilibrium $u^* = (u_1^*, \cdots, u_n^*)$, and for any $\phi \in C$, the solution of (6.4) starting from $\phi$ satisfies*

$$|u_i(t) - u_i^*| \le \| \phi - u^* \| \left[ (1 - \delta_i) e^{-\frac{d_i}{c_i} t} + \delta_i e^{-\frac{\sigma}{2} t} \right]$$

*for all $t \ge 0$ and $(i = 1, \cdots, n)$, where*

$$\begin{cases} \delta_i = \dfrac{2}{2d_i - c_i\sigma} \sqrt{\dfrac{2\alpha l_i}{c_i}} \sum_{j=1}^{n} \left( |a_{ij}| + e^{\frac{\sigma\tau}{2}} |b_{ij}| \right) \\ \alpha = \max_{1 \le i \le n} (l_i c_i) + k\tau \max_{1 \le i \le n} (l_i^2). \end{cases}$$

*Proof*: From Lemma 6.5, the network (6.4) has a unique equilibrium $u^* = (u_1^*, \cdots, u_n^*)$. By coordinate translation $x_i(t) = u_i(t) - u_i^*(i = 1, \cdots, n)$, (6.4) can be written in an equivalent form

$$c_i \frac{dx_i(t)}{dt} = -d_i x_i(t) + \sum_{j=1}^{n} [a_{ij} h_j(x_j(t)) + b_{ij} h_j(x_j(t - \tau_j))] \tag{6.5}$$

where $h_i(x_i) = g_i(x_i + u_i^*) - g_i(u_i^*)$. It's vector form is as follows

$$C \frac{dx(t)}{dt} = -Dx(t) + AH(x(t)) + BH(x(t - \tau)) \tag{6.6}$$

where

$$\begin{cases} H(x(t)) = (h_1(x_1(t)), \cdots, h_n(x_n(t)))^T \\ H(x(t - \tau)) = (h_1(x_1(t - \tau_1)), \cdots, h_n(x_n(t - \tau_n)))^T. \end{cases}$$

By Lemma 6.1 and 6.2, clearly, $h_i(s)(i = 1, \cdots, n)$ are monotonically increasing nonconstant continuous functions with $h_i(0) = 0$ and

$$\begin{cases} h_i^2(s) \leq l_i s h_i(s) \leq l_i^2 s^2 \\ \int_0^s h_i(\theta) d\theta \geq s h_i(s) \end{cases} \tag{6.7}$$

for all $s \in R$ and $(i = 1, \cdots, n)$. And from the Lemma 6.3,

$$\int_0^s h_i(\theta) d\theta \geq \frac{1}{2l_i} h_i^2(s), (i = 1, \cdots, n) \tag{6.8}$$

for all $s \in R$.

Define a differentiable function

$$V_1(t) = e^{\sigma t} \sum_{i=1}^{n} c_i \int_0^{x_i(t)} h_i(\theta) d\theta \tag{6.9}$$

for all $t \geq 0$. From (6.9), (6.5), (6.6) and (6.7), it follows that

$$\begin{aligned} \dot{V}_1(t) &= \sigma V_1(t) + e^{\sigma t} H^T(x(t)) C \dot{x}(t) \\ &= \sigma V_1(t) - e^{\sigma t} H^T(x(t)) D x(t) \\ &\quad + e^{\sigma t} H^T(x(t)) [AH(x(t)) + BH(x(t - \tau))] \\ &\leq \sigma V(t) - \sigma e^{\sigma t} H^T(x(t)) C x(t) \\ &\quad + e^{\sigma t} \left[ -H^T(x(t))(D - \sigma C) x(t) + H^T(x(t)) A H(x(t)) \right] \\ &\quad + e^{\sigma t} H^T(x(t)) B H(x(t - \tau)) \\ &\leq e^{\sigma t} H^T(x(t)) \left[ -(D - \sigma C) L^{-1} + A \right] H(x(t)) \\ &\quad + e^{\sigma t} H^T(x(t)) B H(x(t - \tau)) \end{aligned} \tag{6.10}$$

for all $t \geq 0$.

Define a Lyapunov functional

$$V(t) = V_1(t) + k\sum_{i=1}^{n}\int_{t-\tau_i}^{t} h_i^2(x_i(\theta))e^{\sigma\theta}d\theta \tag{6.11}$$

for all $t \geq 0$. From (6.11), (6.8) and (6.10), it follows that

$$\begin{aligned}
\dot{V}(t) &= \dot{V}_1(t) + ke^{\sigma t}H^T(x(t))H(x(t)) \\
&\quad -ke^{\sigma(t-\tau)}H^T(x(t-\tau))H(x(t-\tau)) \\
&\leq e^{\sigma t}H^T(x(t))\left[-(D-\sigma C)L^{-1} + A + kE\right]H(x(t)) \\
&\quad +e^{\sigma t}H^T(x(t))BH(x(t-\tau)) \\
&\quad - ke^{\sigma(t-\tau)}H^T(x(t-\tau))H(x(t-\tau)) \\
&= e^{\sigma t}H^T(x(t))\Bigg[-(D-\sigma C)L^{-1} + A \\
&\quad +kE + \frac{e^{\sigma\tau}}{4k}BB^T\Bigg]H(x(t)) \\
&\quad +\frac{e^{\sigma\tau}}{4k}e^{\sigma t}H^T(x(t))BB^TH(x(t)) \\
&\quad +e^{\sigma t}H^T(x(t))BH(x(t-\tau)) \\
&\quad -ke^{\sigma(t-\tau)}H^T(x(t-\tau))H(x(t-\tau)) \\
&\leq -\left[\frac{e^{\sigma\tau}}{2k}B^TH(x(t)) - H(x(t-\tau))\right]^T \\
&\quad \times\left[\frac{e^{\sigma\tau}}{2k}B^TH(x(t)) - H(x(t-\tau))\right]ke^{\sigma(t-\tau)} \\
&\leq 0
\end{aligned}$$

for all $t \geq 0$. Thus, $V(t) \leq V(0)$ for all $t \geq 0$.

By (6.11) and (6.8),

$$\begin{aligned}
V(t) &\geq V_1(t) \\
&\geq e^{\sigma t}\sum_{i=1}^{n}c_i\int_0^{x_i(t)} h_i(\theta)d\theta \\
&\geq e^{\sigma t}\sum_{i=1}^{n}\frac{c_i}{2l_i}h_i^2(x_i(t)) \\
&\geq \frac{e^{\sigma t}c_i}{2l_i}h_i^2(x_i(t))
\end{aligned} \tag{6.12}$$

for all $t \geq 0$. In addition,

$$
\begin{aligned}
V(0) &\leq \sum_{i=1}^{n} c_i \int_0^{x_i(0)} h_i(\theta)d\theta + k\sum_{i=1}^{n}\int_{-\tau_i}^{0} h_i^2(x_i(\theta))d\theta \\
&\leq \sum_{i=1}^{n} c_i x_i(0)h_i(x_i(0)) + k\int_{-\tau_i}^{0}\left(\sum_{i=1}^{n} l_i^2 x_i^2(\theta)d\theta\right) \\
&\leq \max_{1\leq i\leq n}(l_i c_i)\sum_{i=1}^{n} x_i^2(0) + k\max_{1\leq i\leq n}(l_i^2)\int_{-\tau}^{0}\sum_{i=1}^{n}|x_i(\theta)|^2 d\theta \\
&\leq \alpha \parallel \phi - u^* \parallel^2 .
\end{aligned} \tag{6.13}
$$

Then, from (6.12) and (6.13),

$$
h_i(x_i(t)) \leq \sqrt{\frac{2\alpha l_i}{c_i}} \parallel \phi - u^* \parallel e^{-\frac{\sigma}{2}t}, (i = 1, \cdots, n) \tag{6.14}
$$

for all $t \geq 0$.

By (6.4) and (6.14), it follows that

$$
\begin{aligned}
c_i D^+|x_i(t)| &\leq -d_i|x_i(t)| + \sum_{j=1}^{n}[|a_{ij}||h_j(x_j(t)| + |b_{ij}||h_j(x_j(t-\tau_j))|] \\
&\leq -d_i|x_i(t)| + \gamma \parallel \phi - u^* \parallel e^{-\frac{\sigma}{2}t}
\end{aligned}
$$

for all $t \geq 0$, where $\gamma = \sqrt{\frac{2\alpha l_i}{c_i}}\sum_{j=1}^{n}\left(|a_{ij}| + e^{\frac{\sigma\tau}{2}}|b_{ij}|\right)$. Then,

$$
\begin{aligned}
|x_i(t)| &\leq e^{-\frac{d_i}{c_i}t}|x_i(0)| + \frac{\gamma}{c_i} \parallel \phi - u^* \parallel \int_0^t e^{-\frac{d_i}{c_i}(t-s)-\frac{\sigma}{2}s}ds \\
&\leq \parallel \phi - u^* \parallel \left[(1-\delta_i)e^{-\frac{d_i}{c_i}t} + \delta_i e^{-\frac{\sigma}{2}t}\right]
\end{aligned}
$$

for all $t \geq 0$. That is

$$
|u_i(t) - u_i^*| \leq \parallel \phi - u^* \parallel \left[(1-\delta_i)e^{-\frac{d_i}{c_i}t} + \delta_i e^{-\frac{\sigma}{2}t}\right]
$$

for all $t \geq 0$. The proof is completed.

In the above theorem, the condition contains two parameters $\sigma$ and $k$. This gives more flexible to the conditions, but additional computations will be needed to select such parameters. The next theorem contains one parameter only.

THEOREM 6.2 *If there exists a positive constant $\sigma$ such that $0 < \sigma \leq d_i/c_i (i = 1, \cdots, n)$ and*

$$
-(D - \sigma C)L^{-1} + \frac{A + A^T}{2} + e^{\frac{\sigma\tau}{2}}\sqrt{\lambda_{max}(BB^T)}E < 0,
$$

*the result of Theorem 6.1 holds.*

*Proof*: Taking

$$k = \frac{e^{\frac{\sigma\tau}{2}}\sqrt{\lambda_{max}(BB^T)}}{2}.$$

Then for any $x \in R^n$, it follows that

$$\begin{aligned} x^T\left[kE + \frac{e^{\sigma\tau}}{4k}BB^T\right]x &\leq kx^Tx + \frac{e^{\sigma\tau}}{4k}\lambda_{max}\left(BB^T\right)x^Tx \\ &= e^{\frac{\sigma\tau}{2}}\sqrt{\lambda_{max}(BB^T)}x^Tx. \end{aligned}$$

Thus, the matrix

$$-(D-\sigma C)L^{-1} + \frac{A+A^T}{2} + kE + \frac{e^{\sigma\tau}}{4k}BB^T$$

is negative definite. Then, Theorem 6.1 is applied and the proof is completed.

## 2.3 Global Exponential Stability

In this section, conditions for GES of (6.4) will be derived.

THEOREM 6.3 *If there exists a constant $k > 0$ such that*

$$-DL^{-1} + \frac{A+A^T}{2} + kE + \frac{1}{4k}BB^T < 0,$$

*the network (6.4) is GES.*

*Proof*: Since

$$-DL^{-1} + \frac{A+A^T}{2} + kE + \frac{1}{4k}BB^T < 0,$$

there must exist $0 < \sigma \leq d_i/c_i (i = 1, \cdots, n)$ such that

$$-(D-\sigma C)L^{-1} + \frac{A+A^T}{2} + kE + \frac{e^{\sigma\tau}}{4k}BB^T < 0.$$

The result now follows from Theorem 6.1. The proof is completed.

THEOREM 6.4 *If*

$$-DL^{-1} + \frac{A+A^T}{2} + \sqrt{\lambda_{max}(BB^T)}E < 0,$$

*the network (6.4) is GES.*

*Proof*: Similar to the proof of Theorem 6.3, under the condition of Theorem 6.4, there must exist a $0 < \sigma \leq d_i/c_i (i = 1, \cdots, n)$ such that

$$-(D - \sigma C)L^{-1} + \frac{A + A^T}{2} + e^{\frac{\sigma \tau}{2}} \sqrt{\lambda_{max} (BB^T)} E < 0.$$

The result now follows from Theorem 6.2 and the proof is completed.

To make the above stability conditions more flexible, some parameters can be introduced.

THEOREM 6.5 *If there exists a positive diagonal matrix $P$ and a constant $k > 0$ such that*

$$-PDL^{-1} + \frac{PA + A^T P}{2} + kE + \frac{1}{4k} PBB^T P < 0,$$

*the network (6.4) is GES.*

*Proof*: Rewrite (6.4) as

$$PC\frac{du(t)}{dt} = -PDu(t) + PAG(u(t)) + PBG(u(t - \tau)) + PI.$$

Then, the the result follows directly from Theorem 6.3. The proof is completed.

## 2.4 Discussions and Illustrative Examples

In [12], the GAS is studied for delayed cellular neural networks described by

$$\dot{x}(t) = -x(t) + Af(x(t)) + Bf(x(t - \tau)) + I$$

where $f(s) = [|s + 1| - |s - 1|]/2$ for $s \in R$ and $\tau > 0$ is a constant. This is a special case of the model (6.4). The GAS condition given by [12] is that

$$\begin{cases} \left(A + A^T\right) \text{ is negative definite} \\ \sqrt{\lambda_{max}(B^T B)} \leq 1. \end{cases}$$

Clearly, the condition given in Theorem 6.4 is weaker than the above one. Moreover, the condition in Theorem 6.4 guarantee GES which is stronger than GAS.

The GAS of RNNs with delays are also studied in [92, 93]. Apply the results in [92, 93] to the network (6.4), the condition for GAS is that

$$\beta(A + A^T) + \beta^2 BB^T - 2\beta DL^{-1} + E < 0,$$

where $\beta = \max_i(l_i/d_i)$. Clearly, this is a special case of the condition for GES in Theorem 6.3 by taking $k = 1/(2\beta)$.

Next, some simple examples will be given to further illustrate the GES conditions.

EXAMPLE 6.1 *Consider a simple one-dimensional RNN with delay*

$$\frac{dx(t)}{dt} = -x(t) + af(x(t)) + bf(x(t-\tau)) \tag{6.15}$$

*where* $f(s) = [|s+1| - |s-1|]/2$ *for all* $s \in R$.

Using Theorem 6.4, if $-1 + a + |b| < 0$, this network is GES. Apply the result in [92, 93] to this network, it requires $-1 + a + (b^2+1)/2 < 0$ for GAS. Obviously, the former condition is weaker. In fact, taking $a = -3$ and $b = 3$, the network is GES by Theorem 6.4, however the stability cannot be checked by [92, 93].

EXAMPLE 6.2 *Consider a two dimensional RNN with delays*

$$\begin{aligned} \begin{bmatrix} \dot{x}_1(t) \\ \dot{x}_2(t) \end{bmatrix} &= -\begin{bmatrix} x_1(t) \\ x_2(t) \end{bmatrix} + \begin{bmatrix} -4 & 3 \\ -3 & -4 \end{bmatrix} \begin{bmatrix} f(x_1(t)) \\ f(x_2(t)) \end{bmatrix} \\ &+ \begin{bmatrix} 2 & 2 \\ 2 & 2 \end{bmatrix} \begin{bmatrix} f(x_1(t-\tau_1)) \\ f(x_2(t-\tau_2)) \end{bmatrix} \end{aligned}$$

*where* $f(s) = [|s+1| - |s-1|]/2$ *for all* $s \in R$ *and* $\tau_1 > 0, \tau_2 > 0$ *are constants.*

Let

$$A = \begin{bmatrix} -4 & 3 \\ -3 & -4 \end{bmatrix}, B = \begin{bmatrix} 2 & 2 \\ 2 & 2 \end{bmatrix}$$

and $D = C = E$. It follows that $L = E$ and $\lambda_{max}\left(BB^T\right) = 16$. Then,

$$-DL^{-1} + \left(A + A^T\right)/2 + \sqrt{\lambda_{max}\left(BB^T\right)}E = \begin{bmatrix} -1 & 0 \\ 0 & -1 \end{bmatrix} < 0.$$

By Theorem 6.4, the network is GES. However, the global stability of this network cannot be checked by the previously known conditions, say, in [8, 10, 12, 29, 68, 69, 118, 92, 93, 158, 159, 160, 224, 225].

## 2.5 Conclusions

In this section, GES of RNNs with constant delays is studied. Global exponential convergence rates are estimated via constructing Lyapunov functionals. New conditions for GES are given. These results generalize the results in [220] for neural networks without delays. Examples are given to show the advantages of these conditions.

One conjecture is that: Can the GES condition in this section be replaced by

$$-DL^{-1} + \frac{A + A^T + B + B^T}{2} < 0$$

or not? If not, does there exist any counter example? It would be an interesting problem to be studied in the future.

Global stability of RNNs with constant delays has been studied by many authors. Readers may refer to [8, 9, 10, 12, 29, 33, 34, 52, 56, 70, 73, 117, 118, 119, 120, 127, 128, 158, 159, 160, 224] for more results.

## 3. RNNs with Variable Delays

### 3.1 Preliminaries

Consider the model of RNNs with variable delays,

$$\begin{cases} C_i \dfrac{du_i(t)}{dt} = -\dfrac{u_i(t)}{R_i} + \sum\limits_{j=1}^{n} \left[ T_{ij} v_j(t) + T_{ij}^{\tau} v_j(t - \tau_{ij}(t)) \right] + I_i \\ v_i(t) = g_i[u_i(t)], (i = 1, \cdots, n) \end{cases} \tag{6.16}$$

for $t \geq 0$, where $C_i > 0, R_i > 0, T_{ij}, T_{ij}^{\tau}$ and $I_i$ are constants, the delays $\tau_{ij}(t)(i = 1, 2, \cdots, n)$ are nonnegative continuous functions with $0 \leq \tau_{ij}(t) \leq \tau$ for $t \geq 0$, where $\tau$ is a constant, the activation functions $g_i \in S_{GLAF}(i = 1, \cdots, n)$, i.e.,

$$0 \leq \frac{g_i(\theta) - g_i(\eta)}{\theta - \eta} \leq l_i, (i = 1, \cdots, n)$$

for all $\theta, \eta \in R$ and $\theta \neq \eta$.

Define a normal in $R^n$ by $|x| = \sum_{i=1}^{n} |x_i|$ for any $x = (x_1, \cdots, x_n)^T \in R^n$. Given any $\phi = (\phi_1, \cdots, \phi_n)^T \in C$, a normal in $C$ is defined by $\parallel \phi \parallel = \sup_{-\tau \leq \theta \leq 0} |\phi(\theta)|$.

For any $\phi = (\phi_1, \cdots, \phi_n)^T \in C$, the initial condition in (6.16) is assumed to be

$$u_i(t) = \phi_i(t), -\tau \leq t \leq 0, i = 1, \cdots, n$$

where $\phi_i(t)(i = 1, \cdots, n)$ are continuous functions. Given any $\phi \in C$, under the assumptions in (6.1), there exists a unique solution of (6.16) starting from $\phi$ [78]. This solution is denoted by $U(t, \phi) = (u_1(t, \phi), \cdots, u_n(t, \phi))^T$.

A vector $U^* = (u_1^*, \cdots, u_n^*)^T \in R^n$ is an equilibrium point of (6.16), if it satisfies,

$$\begin{cases} -\dfrac{u_i^*}{R_i} + \sum\limits_{j=1}^{n} \left( T_{ij} + T_{ij}^{\tau} \right) v_j^* + I_i \equiv 0 \\ v_i^* = g_i(u_i^*), (i = 1, \cdots, n). \end{cases}$$

Obviously, $U^* \in C$. $U^*$ is an equilibrium of (6.16), if and only if $U(t, U^*) = U^*$ for all $t \geq 0$.

DEFINITION 6.5 *The network (6.16) is said to be globally exponentially stable, if there exists a unique equilibrium point $U^* = (u_1^*, \cdots, u_n^*)$, and there exist constants $\epsilon > 0$ and $M \geq 1$ such that for any $\phi \in C$ the following inequality holds,*

$$|U(t, \phi) - U^*| \leq M \parallel \phi - U^* \parallel e^{-\epsilon t}$$

*for all $t \geq 0$.*

DEFINITION 6.6 *The network (6.16) is said to be s-globally exponentially stable, if there exists constants $\epsilon > 0$ and $M \geq 1$, such that for any two $\phi, \psi \in C$ the following inequality is satisfied,*

$$|U(t, \phi) - U(t, \psi)| \leq M \parallel \phi - \psi \parallel e^{-\epsilon t}$$

*for all $t \geq 0$.*

## 3.2 Convergence Analysis

In this subsection, the convergence of (6.16) is studied.

THEOREM 6.6 *If the network (6.16) is s-globally exponentially stable, then the network must be globally exponentially stable.*

*Proof*: Given any $\phi, \psi \in C$, let $U(t, \phi) = [u_1(t, \phi), \cdots, u_n(t, \phi)]^T$ and $U(t, \psi) = [u_1(t, \psi), \cdots, u_n(t, \psi)]^T$ be the solutions of (1) starting from $\phi$ and $\psi$, respectively. Define,

$$\begin{cases} U_t(\phi) = U(t + \theta, \phi) \\ U_t(\psi) = U(t + \theta, \psi), \quad \theta \in [-\tau, 0] \end{cases}$$

for $t \geq 0$. Then, $U_t(\phi), U_t(\psi) \in C$ for all $t \geq 0$.

Since the network (6.16) is s-globally exponentially stable, there exist constants $\epsilon > 0$ and $M \geq 1$ such that

$$|U(t, \phi) - U(t, \psi)| \leq M e^{-\epsilon t} \parallel \phi - \psi \parallel$$

for all $t \geq 0$. Then, it follows for all $t \geq 0$, that

$$\begin{aligned} \parallel U_t(\phi) - U_t(\psi) \parallel &= \sup_{-\tau \leq \theta \leq 0} |U(t + \theta, \phi) - U(t + \theta, \psi)| \\ &\leq e^{\epsilon \tau} M e^{-\epsilon t} \parallel \phi - \psi \parallel . \end{aligned} \tag{6.17}$$

A standard Cauchy argument can be used to prove the convergence of $U_t$.

Select a constant $m > 0$ such that

$$e^{\epsilon\tau} M e^{-\epsilon m} \le \frac{1}{2}. \tag{6.18}$$

For any $\eta > 0$, select a $T > 0$ such that

$$4e^{\epsilon\tau} M e^{-\epsilon t} \|U_m(\phi) - \phi\| \le \eta \tag{6.19}$$

for all $t \ge T$. Given any constant $p > 0$, from (6.17) and (6.18) the following inequality can be derived,

$$\begin{aligned}
\|U_{t+p}(\phi) - U_t(\phi)\| &\le \|U_{t+p}(\phi) - U_{t+p+m}(\phi)\| \\
&\quad + \|U_{t+p+m}(\phi) - U_{t+m}(\phi)\| \\
&\quad + \|U_{t+m}(\phi) - U_t(\phi)\| \\
&= \|U_{t+p}(\phi) - U_{t+p}(U_m(\phi))\| \\
&\quad + \|U_m(U_{t+p}(\phi)) - U_m(U_t(\phi))\| \\
&\quad + \|U_t(U_m(\phi)) - U_t(\phi)\| \\
&\le e^{\epsilon\tau} M e^{-\epsilon(t+p)} \|\phi - U_m(\phi)\| \\
&\quad + e^{\epsilon\tau} M e^{-\epsilon m} \|U_{t+p}(\phi) - U_t(\phi)\| \\
&\quad + e^{\epsilon\tau} M e^{-\epsilon t} \|U_m(\phi) - \phi\| \\
&\le \frac{1}{2} \|U_{t+p}(\phi) - U_t(\phi)\| \\
&\quad + 2e^{\epsilon\tau} M e^{-\epsilon t} \|U_m(\phi) - \phi\|
\end{aligned}$$

for $t \ge 0$. Then, using (6.19), it follows that

$$\|U_{t+p}(\phi) - U_t(\phi)\| \le 4e^{\epsilon\tau} M e^{-\epsilon t} \|U_m(\phi) - \phi\| \le \eta$$

for all $t \ge T$. By the well known *Cauchy convergence principle*, there must exist a $\phi^* \in C$ such that $\lim_{t \to +\infty} U_t(\phi) = \phi^*$. In particular, $\lim_{t \to +\infty} U(t, \phi) = \phi^*(0) \triangleq \phi^{**}$. Obviously, $\phi^{**}$ is an equilibrium point of the network (6.16). That is $U(t, \phi^{**}) = \phi^{**}$ for all $t \ge 0$. Then, it follows that

$$|U(t, \psi) - \phi^{**}| \le M \|\psi - \phi^{**}\| e^{-\epsilon t}$$

for all $t \ge 0$. Clearly, the network (6.16) can not have other equilibria different from $\phi^{**}$. This shows that the network (6.16) is globally exponentially stable. The proof is completed.

Next, the conditions for global exponential stability of the network (6.16) are derived. Initially few differential inequalities are derived, which are used in the proofs of Theorems 6.7 and 6.8.

Given any $\phi, \psi \in C$, let $U(t,\phi)$ and $U(t,\psi)$ be the solutions of (6.16) starting from $\phi$ and $\psi$, respectively. Denote

$$\begin{cases} \bar{u}_i(t) = u_i(t,\phi) - u_i(t,\psi) \\ \bar{v}_i(t) = v_i(t,\phi) - v_i(t,\psi), (i = 1, \cdots, n) \end{cases}$$

for $t \geq -\tau$. Then, it follows from (6.16) that

$$C_i \frac{d\bar{u}_i(t)}{dt} = -\frac{\bar{u}_i(t)}{R_i} + \sum_{j=1}^{n} \left[ T_{ij}\bar{v}_j(t) + T_{ij}^{\tau}\bar{v}_j(t - \tau_{ij}(t)) \right] \tag{6.20}$$

for all $t \geq 0$ and $i = 1, \cdots, n$. Clearly, $sign(\bar{u}_i(t)) \cdot \bar{v}_i(t) \geq 0$ for all $t \geq 0$, then it follows from (6.20) that

$$\begin{aligned} C_i \cdot D^+|\bar{u}_i(t)| &= C_i \cdot sign(\bar{u}_i(t)) \cdot \frac{d\bar{u}_i(t)}{dt} \\ &\leq -\frac{|\bar{u}_i(t)|}{R_i} + T_{ii}|\bar{v}_i(t)| + \sum_{j=1, j\neq i}^{n} |T_{ij}||\bar{v}_j(t)| \\ &\quad + \sum_{j=1}^{n} |T_{ij}^{\tau}||\bar{v}_j(t - \tau_{ij}(t))| \end{aligned} \tag{6.21}$$

for all $t \geq 0$ and $i = 1, \cdots, n$.

THEOREM 6.7 *Denote*

$$\omega_{ij} = \begin{cases} \dfrac{1}{R_i l_i} - T_{ii}^{+} - |T_{ii}^{\tau}|, & i = j \\ -|T_{ij}| - |T_{ij}^{\tau}|, & i \neq j \end{cases}$$

*for* $i, j = 1, \cdots, n$, *where* $T_{ii}^{+} = \max\{0, T_{ii}\}$. *If the matrix* $(\omega_{ij})_{n\times n}$ *is a nonsingular M-matrix, then the network (6.16) is globally exponentially stable.*

*Proof*: Clearly, we have $|\bar{v}_i(t)| \leq l_i|\bar{u}_i(t)|$ for all $t \geq 0$ and $i = 1, \cdots, n$. Then by (6.21), it follows that

$$\begin{aligned} C_i \cdot D^+|\bar{u}_i(t)| &\leq -\frac{|\bar{u}_i(t)|}{R_i} + T_{ii}^{+} l_i|\bar{u}_i(t)| + \sum_{j=1, j\neq i}^{n} l_j|T_{ij}||\bar{u}_j(t)| \\ &\quad + \sum_{j=1}^{n} l_j|T_{ij}^{\tau}||\bar{u}_j(t - \tau_{ij}(t))| \end{aligned} \tag{6.22}$$

for all $t \geq 0$ and $i = 1, \cdots, n$.

Since $(\omega_{ij})_{n\times n}$ is a nonsingular *M-matrix*, there exist constants $\alpha_i > 0 (i = 1, \cdots, n)$ such that $\alpha_i\omega_{ii} + \sum_{j=1, j\neq i}^{n} \alpha_j\omega_{ij} > 0, (i = 1, \cdots, n)$. That is

$$-\frac{1}{R_i l_i} + T_{ii}^{+} + |T_{ii}^{\tau}| + \frac{1}{\alpha_i} \sum_{j=1, j\neq i}^{n} \alpha_j \left( |T_{ij}| + |T_{ij}^{\tau}| \right) < 0$$

for $i = 1, \cdots, n$. Then, a sufficiently small constant $\epsilon > 0$ can be chosen such that

$$\begin{aligned} \xi_i &\triangleq -\frac{1}{R_i l_i} + T_{ii}^+ + \frac{\epsilon}{l_i} + \frac{1}{\alpha_i} \sum_{j=1, j\neq i}^{n} \alpha_j |T_{ij}| \\ &\quad + \frac{e^{\epsilon\tau}}{\alpha_i} \sum_{j=1}^{n} \alpha_j |T_{ij}^{\tau}| \\ &< 0, \quad (i = 1, \cdots, n). \end{aligned}$$

By defining the continuous functions

$$z_i(t) = \frac{l_i}{\alpha_i} |\bar{u}_i(t)| e^{\epsilon t}, \quad (i = 1, \cdots, n)$$

for all $t \geq 0$, it follows from (6.22) that

$$\begin{aligned} C_i \cdot D^+ z_i(t) &\leq \left( -\frac{1}{R_i} + T_{ii}^+ l_i + \epsilon \right) z_i(t) \\ &\quad + \frac{l_i}{\alpha_i} \sum_{j=1, j\neq i}^{n} \alpha_j |T_{ij}| z_j(t) \\ &\quad + \frac{l_i e^{\epsilon\tau}}{\alpha_i} \sum_{j=1}^{n} \alpha_j |T_{ij}^{\tau}| z_j(t - \tau_{ij}(t)) \end{aligned} \tag{6.23}$$

for all $t \geq 0$ and $i = 1, \cdots, n$.

Given any constant $d > 1$, it can be proved that,

$$z_i(t) < d \cdot \max_{1\leq j\leq n} \left\{ \frac{l_j}{\alpha_j} \right\} \cdot \|\phi - \psi\| \triangleq M, (i = 1, \cdots, n) \tag{6.24}$$

for all $t \geq 0$. Otherwise, since $z_i(0) < M, (i = 1, \cdots, n)$, there must exist some $i$ and a $t_1 > 0$ such that $z_i(t_1) = M$ and,

$$z_j(t) \begin{cases} < M, -\tau \leq t < t_1, \text{if } j = i \\ \leq M, -\tau \leq t \leq t_1, \text{if } j \neq i. \end{cases}$$

It follows that $D^+ z_i(t_1) \geq 0$. However, from (6.23)

$$\begin{aligned} C_i \cdot D^+ z_i(t_1) &\leq l_i \Bigg[ -\frac{1}{R_i l_i} + T_{ii}^+ + \frac{\epsilon}{l_i} \\ &\quad + \frac{1}{\alpha_i} \sum_{j=1, j\neq i}^{n} \alpha_j |T_{ij}| + \frac{e^{\epsilon\tau}}{\alpha_i} \sum_{j=1}^{n} \alpha_j |T_{ij}^{\tau}| \Bigg] M \\ &= l_i \cdot \xi_i \cdot M \\ &< 0. \end{aligned}$$

This is a contradiction and thus it proves that (6.24) is true. Letting $d \to 1$ in (6.24), the following inequality is obtained,

$$|\bar{u}_i(t)| \leq \max_{1\leq j\leq n}\left\{\frac{\alpha_j}{l_j}\right\} \cdot \max_{1\leq j\leq n}\left\{\frac{l_j}{\alpha_j}\right\} \cdot \|\phi - \psi\| e^{-\epsilon t}$$

for all $t \geq 0$ and $i = 1, \cdots, n$. This shows the network (6.16) is s-globally exponentially stable. By Theorem 6.6, the network (6.16) is globally exponentially stable.

The stability condition in Theorem 6.7 is independent of the delays. The delays could be any nonnegative continuous bounded functions. Moreover, these delays could be uncertain. These properties provide flexibility in the design of delayed type neural networks. However, since $T_{ii}^{+}$ is used instead of $T_{ii}$ in the stability condition, it is relatively restrictive when $T_{ii} < 0$. A stability condition that fully utilizes each neuron's self connection weight $T_{ii}$ is provided in what follows. However, the stability condition will depend on the delays.

THEOREM 6.8 *Assume the delays are differentiable and satisfy*

$$\frac{d\tau_{ij}(t)}{dt} \leq k_{ij} < 1, (i, j = 1, \cdots, n)$$

*for all* $t \geq 0$*, where* $k_{ij}(i, j = 1, \cdots, n)$ *are constants. Denote*

$$\omega_{ij} = \begin{cases} \dfrac{1}{R_i l_i} - T_{ii} - \dfrac{|T_{ii}^{\tau}|}{1 - k_{ii}}, & i = j \\ -|T_{ij}| - \dfrac{|T_{ij}^{\tau}|}{1 - k_{ij}}, & i \neq j \end{cases}$$

*for* $i, j = 1, \cdots, n$*. If the matrix* $(\omega_{ij})_{n\times n}$ *is a nonsingular M-matrix, then the network (6.16) is globally exponentially stable.*

*Proof*: Since $(\omega_{ij})_{n\times n}$ is a nonsingular *M-matrix*, there exist constants $\alpha_i > 0 (i = 1, \cdots, n)$ such that $\alpha_j \omega_{jj} + \sum_{i=1, i\neq j}^{n} \alpha_i \omega_{ij} > 0, (j = 1, \cdots, n)$. That is,

$$-\frac{\alpha_j}{R_j l_j} + \alpha_j T_{jj} + \sum_{i=1}^{n} \alpha_i \left[ |T_{ij}|(1 - \delta_{ij}) + \frac{|T_{ij}^{\tau}|}{1 - k_{ij}} \right] < 0$$

for $j = 1, \cdots, n$, where

$$\delta_{ij} = \begin{cases} 1, & i = j \\ 0, & i \neq j. \end{cases}$$

Then, a sufficiently small constant $\epsilon > 0$ can be chosen such that,

$$\eta_j \triangleq \left(\epsilon - \frac{1}{R_j}\right) \frac{\alpha_j}{l_j} + \alpha_j T_{jj}$$

$$
\begin{aligned}
&+\sum_{i=1}^{n}\alpha_i\left[|T_{ij}|(1-\delta_{ij})+\frac{e^{\epsilon\tau}|T_{ij}^{\tau}|}{1-k_{ij}}\right]\\
< \ & 0, \quad (j=1,\cdots,n).
\end{aligned}
$$

By defining the continuous functions

$$
\begin{cases} x_i(t)=|\bar{u}_i(t)|e^{\epsilon t} \\ y_i(t)=|\bar{v}_i(t)|e^{\epsilon t}, \quad (i=1,\cdots,n) \end{cases}
$$

for $t \geq 0$, it follows from (6.21) that

$$
\begin{aligned}
C_i\cdot D^{+}x_i(t) \ \leq \ & -\left(\frac{1}{R_i}-\epsilon\right)x_i(t)+T_{ii}y_i(t)\\
&+\sum_{j=1,j\neq i}^{n}|T_{ij}|y_j(t)+\sum_{j=1}^{n}|T_{ij}^{\tau}|y_j(t-\tau_{ij}(t))e^{\epsilon\cdot\tau_{ij}(t)}\\
\leq \ & \left[\left(\epsilon-\frac{1}{R_i}\right)\frac{1}{l_i}+T_{ii}\right]y_i(t)+\sum_{j=1,j\neq i}^{n}|T_{ij}|y_j(t)\\
&+e^{\epsilon\tau}\sum_{j=1}^{n}|T_{ij}^{\tau}|y_j(t-\tau_{ij}(t)) \qquad (6.25)
\end{aligned}
$$

for all $t \geq 0$ and $i = 1,\cdots,n$.

Constructing a Lyapunov functional

$$
V(t)=\sum_{i=1}^{n}\alpha_i\left[C_ix_i(t)+\sum_{j=1}^{n}\frac{|T_{ij}^{\tau}|e^{\epsilon\tau}}{1-k_{ij}}\int_{t-\tau_{ij}(t)}^{t}y_j(s)ds\right] \qquad (6.26)
$$

for all $t \geq 0$.

It follows from (6.25) and (6.26) that

$$
\begin{aligned}
D^{+}V(t) \ \leq \ & \sum_{i=1}^{n}\alpha_i\Bigg[C_i\cdot D^{+}x_i(t)+e^{\epsilon\tau}\sum_{j=1}^{n}\Bigg(\frac{|T_{ij}^{\tau}|}{1-k_{ij}}y_j(t)\\
&-|T_{ij}^{\tau}|y_j(t-\tau_{ij}(t))\Bigg)\Bigg]\\
\leq \ & \sum_{i=1}^{n}\alpha_i\Bigg[\left(\frac{\epsilon}{l_i}-\frac{1}{R_il_i}+T_{ii}\right)y_i(t)\\
&+\sum_{j=1}^{n}\left(|T_{ij}|(1-\delta_{ij})+\frac{e^{\epsilon\tau}|T_{ij}^{\tau}|}{1-k_{ij}}\right)y_j(t)\Bigg]\\
= \ & \sum_{j=1}^{n}\eta_j\cdot y_j(t)\\
\leq \ & 0 \qquad (6.27)
\end{aligned}
$$

for all $t \geq 0$. From (6.27), it follows that

$$V(t) \leq V(0) \tag{6.28}$$

for all $t \geq 0$.

By the definition of $V(t)$ in (6.26),

$$\begin{aligned} V(0) &= \sum_{i=1}^{n} \alpha_i \left[ C_i |\bar{u}_i(0)| \right. \\ &\quad \left. + \sum_{j=1}^{n} \frac{|T_{ij}^{\tau}| e^{\epsilon\tau}}{1-k_{ij}} \int_{-\tau}^{0} |\bar{v}_j(s)| e^{\epsilon s} ds \right] \\ &\leq M_1 \parallel \phi - \psi \parallel \end{aligned}$$

where

$$M_1 = \max_{1 \leq i \leq n} (\alpha_i C_i) + \frac{e^{\epsilon\tau}}{\epsilon} \sum_{i=1}^{n} \alpha_i \max_{1 \leq j \leq n} \left( \frac{|T_{ij}^{\tau}| l_j}{1-k_{ij}} \right),$$

and

$$V(t) \geq e^{\epsilon t} \sum_{i=1}^{n} \alpha_i C_i |\bar{u}_i(t)| \geq \delta e^{\epsilon t} |U(t,\phi) - U(t,\psi)|$$

for all $t \geq 0$, where $\delta = \min_{1 \leq i \leq n}(\alpha_i C_i)$. Then, from (6.28)

$$|U(t,\phi) - U(t,\psi)| \leq M_2 \parallel \phi - \psi \parallel e^{-\epsilon t}$$

for all $t \geq 0$, where $M_2 = \frac{M_1}{\delta} \geq 1$. Thus the network (6.16) is s-globally exponentially stable and by Theorem 6.6, the network (6.16) is globally exponentially stable. The proof is completed.

It is shown that by using the defined Lyapunov functional in (6.26) the conditions for global exponential stability can be derived. This is possible as the Lyapunov functional contains a special term $e^{\epsilon t}$. In addition, this Lyapunov functional can deal with variable delays.

## 3.3 Examples and Discussions

Two examples to illustrate the proposed theory are provided in this section.

EXAMPLE 6.3 *Consider the neural network with variable delays*

$$\begin{cases} \dfrac{du_1(t)}{dt} = -3u_1(t) + g[u_1(t)] - g[u_2(t-\tau_1(t))] + I_1 \\ \dfrac{du_2(t)}{dt} = -2u_2(t) - g[u_2(t)] + g[u_1(t-\tau_2(t))] + I_2 \end{cases} \tag{6.29}$$

*for all $t \geq 0$.*

Where $I_1$ and $I_2$ are constants, $g(s)$ is any continuous function that satisfies,

$$0 \leq \frac{g(\theta) - g(s)}{\theta - s} \leq 1, \quad (i = 1, \cdots, n)$$

for all $\theta, s \in R, \theta \neq s$. The delays $\tau_i(t)(i = 1, 2)$ in the network are assumed to be any nonnegative bounded continuous functions. Moreover, these delays could be uncertain.

It is easy to check that this network satisfies all the conditions of Theorem 6.7 and thus it is globally exponentially stable.

Recently in [92] and [93], neural networks with variable delays were studied and conditions for asymptotic stability were derived. These conditions depend on an upper bound estimation of the derivatives of the delays and, thus can not be used to check the stability of the network (6.29).

EXAMPLE 6.4 *Consider the following neural network with variable delays*

$$\left\{\begin{array}{rcl} \dfrac{du_1(t)}{dt} & = & -u_1(t) - 4g[u_1(t)] + g\left[u_1\left(t - \dfrac{\sin^2 t}{2}\right)\right] \\ & & -g\left[u_2\left(t - \dfrac{\cos^2 t}{2}\right)\right] + I_1 \\ \dfrac{du_2(t)}{dt} & = & -u_2(t) - 2g[u_2(t)] - g\left[u_1\left(t - \dfrac{\sin^2 t}{4}\right)\right] \\ & & +g\left[u_2\left(t - \dfrac{\cos^2 t}{4}\right)\right] + I_2 \end{array}\right. \tag{6.30}$$

*for all* $t \geq 0$.

Where $I_1$ and $I_2$ are constants, and $g(s)$ is defined by

$$g(s) = \frac{|s+1| - |s-1|}{2}, \quad s \in R.$$

It is noted that the network (6.30) does not satisfy the conditions of Theorem 6.7 and therefore its stability can not be checked by Theorem 6.7. Next, Theorem 6.8 is used to check the stability.

In network (6.30), $R_1 = R_2 = C_1 = C_2 = 1$, $T_{11} = -4$, $T_{12} = T_{21} = 0$, $T_{22} = -2$, $T_{11}^{\tau} = T_{22}^{\tau} = 1$, $T_{12}^{\tau} = T_{21}^{\tau} = -1$, $\tau_{11}(t) = \sin^2 t/2$, $\tau_{12}(t) = \cos^2 t/2$, $\tau_{21}(t) = \sin^2 t/4$, $\tau_{22}(t) = \cos^2 t/4$. By the definition of $g(s)$, $k_1 = k_2 = 1$. Since the delays are differentiable, then

$$\frac{d\tau_{11}(t)}{dt} = \frac{\sin 2t}{2} \leq \frac{1}{2} \triangleq l_{11} < 1$$

for all $t \geq 0$. Similarly, $k_{12} = 1/2$, $k_{21} = k_{22} = 1/4$. Then, it follows that

$$\frac{1}{R_1 l_1} - T_{11} - \frac{|T_{11}^{\tau}|}{1 - k_{11}} - |T_{12}| - \frac{|T_{12}^{\tau}|}{1 - k_{12}} = 1 > 0$$

and

$$\frac{1}{R_2 l_2} - T_{22} - \frac{|T_{22}^{\tau}|}{1 - k_{22}} - |T_{21}| - \frac{|T_{21}^{\tau}|}{1 - k_{21}} = \frac{1}{3} > 0.$$

This shows that the condition in Theorem 6.8 is satisfied. Thus, the network (6.30) is globally exponentially stable.

The network (6.30) also does not satisfy the conditions in [92] or [93]. Moreover, the conditions in [92] and [93] only guarantee *global asymptotic stability*, which is obviously weaker than the *global exponential stability*.

### 3.4 Conclusions

The GES of a class of RNNs with variable delays is studied in this section. The concept of s-global exponential stability is proposed. Using the method of delayed inequality analysis and the method of constructing Lyapunov functional together with the proposed concept of s-global exponential stability, simple conditions for global exponential stability are derived. The condition in Theorem 6.7 is independent of the delays and thus it provides flexibility in the design of neural networks with variable delays. If in the condition of Theorem 6.7, $T_{ii}^{+}(i = 1, \cdots, n)$ can be replaced with $T_{ii}(i = 1, \cdots, n)$, it will provide a better condition for global exponential convergence. More research in this direction is required.

Comparing with RNNs with constant delays, RNNs with variable delays have not received extensive study. For more results on global stability of RNNs with variable delays, see [225, 218, 92, 93, 152, 222].

## 4. Special RNN model with Delays

### 4.1 Preliminaries

The model of RNNs with delays which will be discussed is described by

$$\begin{aligned}\frac{du_i(t)}{dt} &= \sum_{j=1}^{i} \left[a_{ij} g_i(u_j(t)) + b_{ij} g_j(u_j(t - \tau_{ij}(t))\right] \\ &\quad -d_i \cdot u_i(t) + I_i, \quad (i = 1, \cdots, n) \qquad (6.31)\end{aligned}$$

for $t \geq 0$, where

| | |
|---|---|
| $u_i(t)$ | neural voltages; |
| $d_i > 0$ | resistances; |
| $A = (a_{ij})_{n\times n}, B = (b_{ij})_{n\times n}$ | connection matrices; |
| $g_i$ | activation functions; |
| $I_i$ | external inputs. |
| $\tau_{ij}(t)$ | delays. |

Clearly, the connection matrices $A$ and $B$ are triangle matrices.

The delays in the network are assumed to be any nonnegative bounded continuous functions. Thus, there exists a constant $\tau \geq 0$ such that $0 \leq \tau_{ij}(t) \leq \tau, \quad (i,j = 1,\cdots,n)$ for all $t \geq 0$.

It is assumed that the activation functions $g_i \in S_{GLAF}(i = 1,\cdots,n)$, i.e., there exist constants $l_i(i = 1,\cdots,n)$ such that

$$0 \leq \frac{g_i(\theta) - g_i(\eta)}{\theta - \eta} \leq l_i, (i = 1,\cdots,n)$$

for $\theta, \eta \in R$ and $\theta \neq \eta$.

Let $C = C([-\tau, 0], R^n)$ be the *Banach* space of continuous functions mapping the interval $[-\tau, 0]$ into $R^n$ with the topology of uniform convergence. For any $\phi = (\phi_1,\cdots,\phi_n)^T \in C$, the normal in $C$ is defined by

$$\| \phi \| = \sup_{-\tau \leq \theta \leq 0} |\phi(\theta)|$$

where $|\phi(\theta)| = \sum_{i=1}^{n} |\phi_i(\theta)|$ is a normal in $R^n$.

For any $\phi = (\phi_1,\cdots,\phi_n)^T \in C$, the initial condition of (6.31) is assumed to be

$$x_i(t) = \phi_i(t), \quad -\tau \leq t \leq 0, \quad (i = 1,\cdots,n)$$

where $\phi_i(t)(i = 1,\cdots,n)$ are continuous functions. For any $\phi \in C$, by the condition that $g_i \in S_{GLAF}(i = 1,\cdots,n)$, it follows from [78] that there exists a unique solution of (6.31) starting from $\phi$. This solution is denoted by $U(t,\phi) = (u_1(t,\phi),\cdots,u_n(t,\phi))^T$.

A point $U^* = (u_1^*,\cdots,u_n^*)^T \in R^n$ is an equilibrium point of (6.31) if and only if

$$-d_i \cdot u_i^* + \sum_{j=1}^{i} (a_{ij} + b_{ij})\, g(u_j^*) + I_i \equiv 0, (i = 1,\cdots,n).$$

The network (6.31) is GES, if the network has a unique equilibrium $U^* = (u_1^*,\cdots,u_n^*)$, and there exist constants $\epsilon > 0$ and $\Omega \geq 1$ such that for any $\phi \in C$ it holds that

$$|U(t,\phi) - U^*| \leq \Omega \| \phi - U^* \| e^{-\epsilon t}$$

for all $t \geq 0$.

For any $x \in R$, denote

$$x^+ = \begin{cases} x & \text{if } x \geq 0 \\ 0 & \text{if } x < 0. \end{cases}$$

## 4.2 Equilibrium Point and Convergence Analysis

THEOREM 6.9 *If*

$$-d_i + \left(a_{ii} + b_{ii}\right)^+ l_i < 0, \quad (i = 1, \cdots, n),$$

*there exists a unique equilibrium point of the network (6.31).*

*Proof*: Initially it is proved that there exists an equilibrium point $(u_1^*, \cdots, u_n^*)^n \in R^n$ for the network (6.31) such that

$$-d_i u_i^* + \sum_{j=1}^{i} \left(a_{ij} + b_{ij}\right) g_j(u_j^*) + I_i = 0 \tag{6.32}$$

for $i = 1, \cdots, n$.

Define the following continuous function

$$f_1(\theta) = -d_1 \cdot \theta + \left(a_{11} + b_{11}\right) \cdot g_1(\theta) + I_1$$

for all $\theta \in R$.

From (6.31),

$$\begin{aligned} & -d_1 + \left(a_{11} + b_{11}\right) \frac{g_1(\theta) - g_1(0)}{\theta} \\ \leq \; & \max\left\{-d_1, -d_1 + \left(a_{11} + b_{11}\right)^+ l_1\right\} \\ < \; & 0 \end{aligned}$$

Then, it follows that

$$\begin{aligned} f_1(\theta) &= \left[-d_1 + \left(a_{11} + b_{11}\right) \frac{g_1(\theta) - g_1(0)}{\theta}\right] \theta \\ &\quad + \left(a_{11} + b_{11}\right) g_1(0) + I_1 \\ &= \begin{cases} +\infty, & \theta \to -\infty \\ -\infty, & \theta \to +\infty. \end{cases} \end{aligned}$$

By continuity, there must exist a point $u_1^* \in R$ such that $f_1(u_1^*) = 0$. That is

$$-d_1 u_1^* + \left(a_{11} + b_{11}\right) g_1(x_1^*) + I_1 = 0.$$

This shows that (6.32) holds for $i = 1$.

Suppose that (6.32) holds for $i = 1, \cdots, k-1$, i.e., there exist $u_1^*, \cdots, u_{k-1}^*$ such that

$$-d_i u_i^* + \sum_{j=1}^{i} \left(a_{ij} + b_{ij}\right) g_j(u_j^*) + I_i = 0$$

for $i = 1, \cdots, k-1$. In what follows it is proved that (6.32) is also true for $i = k$.

Define a continuous function

$$\begin{aligned} f_k(\theta) &= -d_k\theta + \left(a_{kk} + b_{kk}\right) \cdot g_k(\theta) \\ &\quad + \sum_{j=1}^{k-1} \left(a_{ij} + b_{ij}\right) g_j(x_j^*) + I_k \end{aligned}$$

for $\theta \in R$. On the lines of the above discussion, there exists a $u_k^* \in R$ such that $f(u_k^*) = 0$, i.e.,

$$-d_k u_k^* + \sum_{j=1}^{k} \left(a_{ij} + b_{ij}\right) g_j(u_j^*) + I_k = 0.$$

This shows that (6.32) also holds for $i = k$. By mathematical deduction, (6.32) holds for all $i = 1, \cdots, n$.

Suppose that the network (6.31) has another equilibrium point $(u_1^\dagger, \cdots, u_n^\dagger)^T \in R^n$, then

$$-d_i u_i^\dagger + \sum_{j=1}^{i} \left(a_{ij} + b_{ij}\right) g(u_j^\dagger) + I_i = 0, (i = 1, \cdots, n). \tag{6.33}$$

From (6.32) and (6.33), it follows that

$$-d_i \left[u_i^* - u_i^\dagger\right] + \sum_{j=1}^{i} \left(a_{ij} + b_{ij}\right) \left[g_j(u_j^*) - g_j(u_j^\dagger)\right] = 0 \tag{6.34}$$

for $i = 1, \cdots, n$, and then

$$-d_1 \left[u_1^* - u_1^\dagger\right] + \left(a_{11} + b_{11}\right) \left[g_1(u_1^*) - g_1(u_1^\dagger)\right] = 0$$

and hence the following inequality holds,

$$\left|u_1^* - u_1^\dagger\right| \left[-d_1 + \left(a_{11} + b_{11}\right)^+ l_1\right] \leq 0$$

The above inequality implies that $u_1^* = u_1^\dagger$. With $u_1^* = u_1^\dagger$ and from (6.34),

$$-d_2\left[u_2^* - u_2^\dagger\right] + \left(a_{22} + b_{22}\right)\left[g_2(u_2^*) - g(u_2^\dagger)\right] = 0.$$

Similarly, it can be proved that $u_2^* = u_2^\dagger$ and in turn $u_i^* = u_i^\dagger$ for all $i = 1, \cdots, n$.

THEOREM 6.10 *If*

$$-d_i + \left(a_{ii}^+ + |b_{ii}|\right)l_i < 0, \quad (i = 1, \cdots, n),$$

*the network (6.31) is GES.*

*Proof*: Since

$$-d_i + \left(a_{ii}^+ + |b_{ii}|\right)l_i < 0, \quad (i = 1, \cdots, n)$$

imply that

$$-d_i + \left(a_{ii} + b_{ii}\right)^+ l_i < 0, \quad (i = 1, \cdots, n)$$

then, by the Theorem 6.9, the network (6.31) has a unique equilibrium point $u^* = (u_1^*, \cdots, u_n^*)^T \in R^n$.

Clearly, there must exist a constant $\epsilon > 0$ such that

$$\epsilon - d_i + a_{ii}^+ l_i + e^{\epsilon\tau}|b_{ii}|l_i < 0, (i = 1, \cdots, n).$$

Denote

$$\begin{cases} \eta = \max\limits_{1\le i\le n}\left\{\dfrac{e^{\epsilon\tau}|b_{ii}|l_i}{d_i - a_{ii}^+ l_i - \epsilon}\right\} \\ M = \max\limits_{1\le i\le n}\left\{\sum\limits_{j=1}^{i-1} l_j\left(|a_{ij}| + e^{\epsilon\tau}|b_{ij}|\right)\right\} \\ \Pi = M \cdot \max\limits_{1\le i\le n}\left\{\dfrac{1}{d_i - a_{ii}^+ l_i - \epsilon}\right\}. \end{cases}$$

Obviously, $\eta < 1$.

Given any $\phi \in C$, let $(u_i(t), \cdots, u_n(t))^T$ be the solution of (6.31) starting from $\phi$.

Initially the following inequality is proved,

$$|u_i(t) - u_i^*| \le \left[1 + \sum_{j=1}^{i-1}\left(\frac{\Pi M}{1-\eta}\right)^{i-1}\right]\frac{\|\phi - u^*\|}{1-\eta}e^{-\epsilon t} \qquad (6.35)$$

for $i = 1, \cdots, n$ and all $t \ge 0$.

It follows from (6.31) that

$$\begin{aligned}\frac{d[u_1(t)-u_1^*]}{dt} &= -d_1[u_1(t)-u_1^*]+a_{11}[g(u_1(t))-g(u_1^*)]\\ &\quad +b_{11}[g(u_1(t-\tau_{ij}(t)))-g(u_1^*)]\end{aligned}$$

for $t \geq 0$. Then,

$$\begin{aligned}D^+|u_1(t)-u_1^*| &= -d_1|u_1(t)-u_1^*|\\ &\quad +a_{11}^+|g(u_1(t))-g(u_1^*)|\\ &\quad +|b_{11}|\,|g(u_1(t-\tau_{11}(t)))-g(u_1^*)|\\ &\leq -(d_1-a_{11}^+l_1)|u_1(t)-u_1^*|\\ &\quad +|b_{11}|\,l_1\,|u_1(t-\tau_{11}(t))-u_1^*| \qquad (6.36)\end{aligned}$$

for $t \geq 0$.

Define

$$z_1(t)=|u_1(t)-u_1^*|e^{\epsilon t}$$

for $t \geq -\tau$. Then, it follows from (6.36) that

$$\begin{aligned}D^+z_1(t) &\leq (\epsilon-d_1+a_{11}^+l_1)z_1(t)\\ &\quad +e^{\tau\epsilon}|b_{11}|l_1z_1(t-\tau_{11}(t))\end{aligned}$$

for all $t \geq 0$, and hence

$$\begin{aligned}z_1(t) &\leq z_1(0)e^{(\epsilon-d_1+a_{11}^+l_1)t}+e^{\tau\epsilon}|b_{11}|l_1\\ &\quad \times\int_0^t e^{(\epsilon-d_1+a_{11}^+l_1)(t-s)}z_1(s-\tau_{11}(s))ds\\ &\leq \|\phi-u^*\|+\eta\cdot\sup_{-\tau\leq s\leq t}\{z_1(s)\} \qquad (6.37)\end{aligned}$$

for all $t \geq 0$. It is easy to note that $z_1(t) \leq \|\phi-u^*\|$ for $-\tau \leq t \leq 0$, then together with (6.37), it follows that

$$\sup_{-\tau\leq s\leq t}\{z_1(s)\}\leq\|\phi-u^*\|+\eta\cdot\sup_{-\tau\leq s\leq t}\{z_1(s)\}$$

for all $t \geq -\tau$. Hence,

$$|u_1(t)-u_1^*|\leq\frac{1}{1-\eta}\|\phi-u^*\|e^{-\epsilon t}$$

for all $t \geq 0$. This shows that (6.35) is true for $i=1$.

Suppose (6.35) is true for $i = 1, \cdots, k-1$, then,

$$
\begin{aligned}
|u_i(t) - u_i^*| &\leq \left[1 + \sum_{j=1}^{i-1} \left(\frac{\Pi M}{1-\eta}\right)^j\right] \frac{\|\phi - u^*\|}{1-\eta} e^{-\epsilon t} \\
&\leq \left[1 + \sum_{j=1}^{k-2} \left(\frac{\Pi M}{1-\eta}\right)^j\right] \frac{\|\phi - u^*\|}{1-\eta} e^{-\epsilon t} \qquad (6.38)
\end{aligned}
$$

for $i = 1, \cdots, k-1$ and all $t \geq 0$.

From (6.31),

$$
\begin{aligned}
\frac{d[u_k(t) - u_k^*]}{dt} &= -d_k[u_k(t) - u_k^*] \\
&\quad + a_{kk}\Big[g(u_k(t)) - g(u_k^*]\Big] \\
&\quad + b_{kk}\Big[g(u_k(t - \tau_{kk}(t))) - g(u_k^*]\Big] \\
&\quad + \sum_{j=1}^{k-1} \Bigg[ a_{ij}\Big[g_j(u_j(t)) - g_j(u_j^*)\Big] \\
&\quad + b_{ij}\Big[g_j(u_j(t - \tau_{kj}(t))) - g_j(u_j^*)\Big]\Bigg]
\end{aligned}
$$

for $t \geq 0$, and then from (6.38),

$$
\begin{aligned}
D^+|u_k(t) - u_k^*| &\leq -d_k\,|u_k(t) - u_k^*| \\
&\quad + a_{kk}^+\,|g(u_k(t)) - g(u_k^*)| \\
&\quad + |b_{kk}|\,|g(u_k(t - \tau_{kk}(t))) - g(u_k^*)| \\
&\quad + \sum_{j=1}^{k-1} \Bigg[ |a_{ij}|\,\Big|g_j(u_j(t)) - g_j(u_j^*)\Big| \\
&\quad + |b_{ij}|\,\Big|g_j(u_j(t - \tau_{kj}(t))) - g_j(u_j^*)\Big|\Bigg] \\
&\leq -\Big(d_k - a_{kk}^+ l_k\Big)\,|u_k(t) - u_k^*| \\
&\quad + |b_{kk}|\,l_k\,|u_k(t - \tau_{kk}(t)) - u_k^*| \\
&\quad + \sum_{j=1}^{k-1} l_j \Bigg[ |a_{kj}|\,\Big|u_j(t) - u_j^*\Big| \\
&\quad + |b_{kj}|\,\Big|u_j(t - \tau_{kj}(t)) - u_j^*\Big|\Bigg] \\
&\leq -\Big(d_k - a_{kk}^+ l_k\Big)\,|u_k(t) - u_k^*|
\end{aligned}
$$

$$+\left|b_{kk}\right| l_k \left|u_k(t-\tau_{kk}(t)) - u_k^*\right| + M\left[1+\sum_{j=1}^{k-2}\left(\frac{\Pi M}{1-\eta}\right)^j\right] \times \frac{\|\phi - u^*\|}{1-\eta} e^{-\epsilon t}$$

for all $t \geq 0$. Define

$$z_k(t) = \left|u_k(t) - u_k^*\right| e^{\epsilon t}$$

for $t \geq -\tau$. It follows that

$$\begin{aligned} D^+ z_k(t) &\leq -\left(d_k - a_{kk}^+ l_k\right) z_k(t) \\ &\quad + e^{\tau\epsilon}\left|b_{kk}\right| l_k z_k(t-\tau_{kk}(t)) \\ &\quad M\left[1+\sum_{j=1}^{k-1}\left(\frac{\Pi M}{1-\eta}\right)^{k-1}\right]\frac{\|\phi - u^*\|}{1-\eta} \end{aligned}$$

for $t \geq 0$, and hence

$$\begin{aligned} z_k(t) &\leq z_k(0)e^{-\left(d_k - a_{kk}^+ l_k\right)} + e^{\tau\delta}\left|b_{kk}\right| l_k \\ &\quad \times \int_0^t e^{-\left(d_k - a_{kk}^+ l_k\right)(t-s)} z_k(s-\tau_{kk}(s))ds \\ &\quad + \Pi M\left[1+\sum_{j=1}^{k-1}\left(\frac{\Pi M}{1-\eta}\right)^{j-1}\right]\frac{\|\phi - u^*\|}{1-\eta} \\ &\leq \left[1+\sum_{j=1}^{k}\left(\frac{\Pi M}{1-\eta}\right)^{j}\right]\|\phi - u^*\| \\ &\quad + \eta \cdot \sup_{-\tau \leq s \leq t}\{z_k(s)\} \end{aligned}$$

for $t \geq 0$. Using the fact that $z_k(t) \leq \|\phi - u^*\|$ for $-\tau \leq t \leq 0$, it follows that

$$\begin{aligned} \sup_{-\tau \leq s \leq t}\{z_k(s)\} &\leq \left[1+\sum_{j=1}^{k-1}\left(\frac{\Pi M}{1-\eta}\right)^{j}\right]\|\phi - u^*\| \\ &\quad + \eta \cdot \sup_{-\tau \leq s \leq t}\{z_k(s)\} \end{aligned}$$

for all $t \geq -\tau$. This means that

$$\sup_{-\tau \leq s \leq t}\{z_k(s)\} \leq \left[1+\sum_{j=1}^{k-1}\left(\frac{\Pi M}{1-\eta}\right)^{j}\right]\frac{\|\phi - u^*\|}{1-\eta}$$

for all $t \geq -\tau$. Then,

$$|u_k(t) - u_k^*| \leq \left[1 + \sum_{j=1}^{k-1} \left(\frac{\Pi M}{1-\eta}\right)^j\right] \frac{\|\phi - u^*\|}{1-\eta} e^{-\epsilon t}$$

for all $t \geq 0$. This proves that (6.35) is true for $i = k$. By mathematical induction, (6.35) holds for all $i = 1, \cdots, n$. Clearly, (6.35) implies that the network (6.31) is GES.

The stability condition given in Theorem 6.10 is very simple and quite easy to verify. The derived condition depends only on the network's resistances and each neuron's self interaction. Since the condition does not depend on the delays, different neurons' interactions, and the special form of the neural activation functions, it provides flexibility in the design of the network.

## 4.3 Examples and Discussions

In this subsection, some examples are provided to illustrate the developed theory.

EXAMPLE 6.5 *Consider the following RNN without any delay*

$$\begin{cases} \dfrac{du_1(t)}{dt} = -2u_1(t) + g_1[u_1(t)] + I_1 \\ \dfrac{du_2(t)}{dt} = -\dfrac{1}{2}u_2(t) + a \cdot g_1[u_1(t)] - \dfrac{1}{2}g_2[u_2(t)] + I_2 \end{cases} \tag{6.39}$$

*where $a \in R$ is a constant, $I_i (i = 1, 2)$ are constants, and $g_i(\cdot)(i = 1, 2)$ are any continuous functions which satisfy*

$$0 \leq \frac{g_i(\theta) - g_i(s)}{\theta - s} \leq 1, (i = 1, 2).$$

For any constant $a \in R$, it is easy to note that the network (6.39) satisfies the stability condition of Theorem 6.10, then it is globally exponentially stable. However, the stability of this network can not be verified by the conditions recently reported in [14, 35, 36, 57, 60, 114, 156, 218]. In fact, with $a = 10$ in the network (6.39), the network does not satisfy any stability conditions in [14, 35, 36, 57, 60, 114, 156, 225].

Figure 6.1 shows the convergence of the network for $a = 10$, $g_1(s) = g_2(s) = 1/(1 + e^{-s})$ and $I_1 = I_2 = 1$.

EXAMPLE 6.6 *Consider the following Cellular Neural Network with delay*

$$\frac{d}{dt}\begin{bmatrix} u_1(t) \\ u_2(t) \\ u_3(t) \end{bmatrix} = -\begin{bmatrix} u_1(t) \\ u_2(t) \\ u_3(t) \end{bmatrix} + \begin{bmatrix} I_1 \\ I_2 \\ I_3 \end{bmatrix}$$

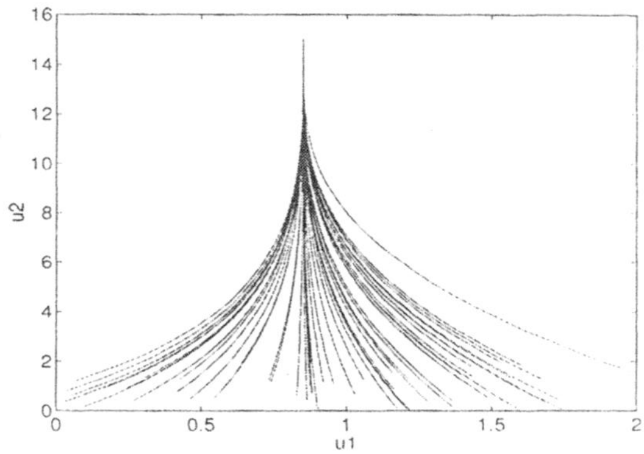


*Figure 6.1.* Convergence of network in (6.39)

$$+\begin{bmatrix} a & b & 0.5 \\ -0.6 & 0 & 0 \\ c & 0.8 & 0 \end{bmatrix}\begin{bmatrix} f(u_1(t)) \\ f(u_2(t)) \\ f(u_3(t)) \end{bmatrix}$$
$$+\begin{bmatrix} d & e & 0.4 \\ 1.5 & 0 & 0 \\ g & 0.1 & 0 \end{bmatrix}\begin{bmatrix} f(u_1(t-\tau_1(t))) \\ f(u_2(t-\tau_2(t))) \\ f(u_3(t-\tau_3(t))) \end{bmatrix} \quad (6.40)$$

*where the delays $\tau_i(t)(i = 1, \cdots, n)$ are any nonnegative bounded continuous functions, and*

$$f(\theta) = \frac{|\theta + 1| - |\theta - 1|}{2}, \theta \in R.$$

This network can also be represented as

$$\begin{bmatrix} \dot{u}_2(t) \\ \dot{u}_3(t) \\ \dot{u}_1(t) \end{bmatrix} = -\begin{bmatrix} u_2(t) \\ u_3(t) \\ u_1(t) \end{bmatrix} + \begin{bmatrix} I_2 \\ I_3 \\ I_1 \end{bmatrix}$$
$$+\begin{bmatrix} -0.6 & 0 & 0 \\ c & 0.8 & 0 \\ a & b & 0.5 \end{bmatrix}\begin{bmatrix} f(u_2(t)) \\ f(u_3(t)) \\ f(u_1(t)) \end{bmatrix}$$
$$+\begin{bmatrix} 0.9 & 0 & 0 \\ d & -0.1 & 0 \\ e & g & 0.4 \end{bmatrix}\begin{bmatrix} f(u_2(t-\tau_2(t))) \\ f(u_3(t-\tau_3(t))) \\ f(u_1(t-\tau_1(t))) \end{bmatrix}$$

Using Theorem 6.10, it is easy to verify that this network is globally exponentially stable subject to any nonnegative delays $\tau_i(t)(i = 1, 2, 3)$, any constants inputs $I_i(i = 1, 2, 3)$, any constants $a, b, c, d, e$ and $g$. However, the stability of this network can not be verified the conditions provided in [8, 10, 12, 68, 69, 92, 93, 119, 118, 127, 159, 224, 225, 218, 231].

Figure 6.2 shows the convergence of the network when $a = 0.9618$, $b = 0.1630$, $c = 0.7486$, $d = 0.3741$, $e = 0.4542$, $g = 0.0386$, $I_1 = I_2 = I_3 = 1$ and $\tau_1(t) = \tau_2(t) = \tau_3(t) = 0$.

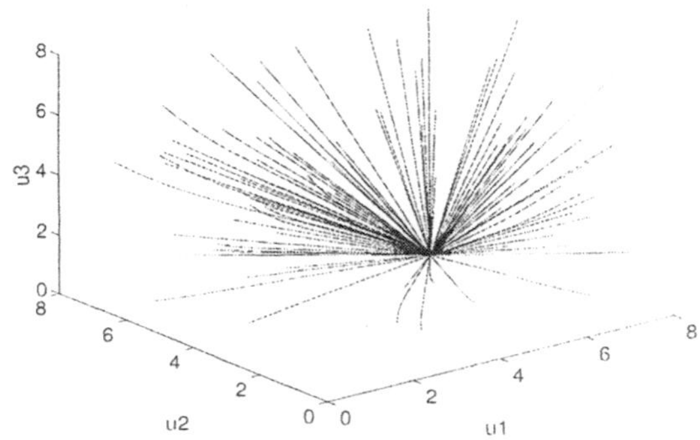


*Figure 6.2.* Convergence of network in (6.40)

## 4.4 Conclusions

In this section, simple conditions for a class of RNNs with delays are derived. It is assumed that the connection matrices of the model are to be in the form of triangle matrices. In some applications, there do exist RNNs that could be described by the models mentioned. Moreover, the conditions for stability can not be included in the known existing conditions for stability of RNNs. This also shows the constriction of the existing stability conditions for RNNs. Further research is required.

# 5. Absolute Stability and Absolute Periodicity

Absolute stability of RNNs without delays has been studied by many authors in recent years, see for examples [10, 14, 61, 99, 100, 116, 113]. The absolute stability of a RNN means that for every activation function in some functional set and every constant input vector, an equilibrium of the network exists and all other solutions of the network converge to this equilibrium. In general, activation functions are nonlinear functions, which makes the corresponding neural networks become nonlinear differential equations. Since an absolutely stable neural network allows its activation functions together with its input vectors to change in some range, it gives flexibility to the design of global stable neural networks. RNNs with the property of absolute stability have found to be very useful in solving some optimization problems.

RNNs with delays have also attracted the interests of scientists, see for examples [8, 12, 44, 67, 68, 69, 137, 159, 179, 191, 32, 224, 225, 218]. It is known that delays can affect the dynamic behavior of neural networks [17, 137]. RNNs with delays have found many applications in processing of moving images, image compression [191], pattern classifications and quadratic programming problems [37]. However, the absolute stability of RNNs with delays has not been widely studied as that of RNNs without delays. Fewer results on absolute stability of RNNs with delays have been reported in [92] recently.

In the application of RNNs to some practical problems, the properties of equilibrium points play important roles. An equilibrium point can be looked as a special periodic solution of RNNs with arbitrary period. In this sense the analysis of periodic solutions of RNNs could be more general than that of equilibrium points. In [224, 225], the existence of periodic solutions together with global exponential stability of RNNs with delays are studied. The existence of periodic solutions of RNNs is an interesting dynamic behaviour. It has been found applications in learning theory [188], which is motivated by the fact that most learning systems need repetition. It is also expected that it can be applied to association memory by storing targets in periodic solutions. It is believed that more applications of such dynamic property of RNNs will be found in the future.

In this section, the absolute periodicity of RNNs with delays is proposed to study. By absolute periodicity of a RNN, it means that for every activation function in some functional set and every input periodic vector function, a unique periodic solution of the network exists and all other solutions converge exponentially to it. The absolute stability of RNNs with delays will be also studied in this section. By developing the methods in [224, 225], some simple and easily checkable conditions will be derived to guarantee the absolute periodicity and absolute stability of RNNs with delays.

The rest of this section is organized as follows. Preliminaries will be given in Subsection 5.1. Main results and proofs will be presented and discussed in Subsection 5.2. Simulation results for absolute periodicity will be given in Subsection 5.3. Finally in Subsection 5.4, a conclusion follows.

## 5.1 Preliminaries

A vector function $I(t) = (I_1(t), \cdots, I_n(t))^T$ is said to be in a set $P$, if $I_i(t)(i = 1, \cdots, n)$ are continuous functions and there exists a constant $\varpi > 0$ such that $I_i(t + \varpi) = I_i(t)(i = 1, \cdots, n)$ for all $t \geq 0$. That is $I(t)$ is a periodic vector function with period $\varpi$.

Suppose $g_i \in S_{GLAF}(i = 1, \cdots, n)$, i.e.,

$$0 \leq \frac{g_i(\theta) - g_i(\eta)}{\theta - \eta} \leq l_i, (i = 1, \cdots, n)$$

for all $\theta, \eta \in R$ and $\theta \neq \eta$. Given any $I(t) \in P$, consider the following RNN with delays

$$\frac{du_i(t)}{dt} = \sum_{j=1}^{n} \left[a_{ij}v_j(t) + b_{ij}v_j(t - \tau_{ij})\right] - d_i u_i(t) + I_i(t), (i = 1, \cdots, n) \tag{6.41}$$

with

$$v_i(t) = g_i[u_i(t)], (i = 1, \cdots, n)$$

for $t \geq 0$, where $d_i > 0 (i = 1, \cdots, n)$ and $\tau_{ij} \geq 0 (i, j = 1, \cdots, n)$.

Given any constant input vector $I = (I_1, \cdots, I_n)^T \in R^n$, consider the following RNN with delays

$$\frac{du_i(t)}{dt} = \sum_{j=1}^{n} \left[a_{ij}v_j(t) + b_{ij}v_j(t - \tau_{ij})\right] - d_i u_i(t) + I_i, (i = 1, \cdots, n) \tag{6.42}$$

with

$$v_i(t) = g_i[u_i(t)], (i = 1, \cdots, n)$$

for $t \geq 0$, where $d_i > 0 (i = 1, \cdots, n)$ and $\tau_{ij} \geq 0 (i, j = 1, \cdots, n)$.

Denote $\tau = \max_{1 \leq i,j \leq n}\{\tau_{ij}\}$. Let $C = C([-\tau, 0], R^n)$ be the *Banach* space of continuous functions mapping the interval $[-\tau, 0]$ into $R^n$ with the topology of uniform convergence. For any $\phi = (\phi_1, \cdots, \phi_n)^T \in C$, define a normal in $C$ by

$$\| \phi \| = \sup_{-\tau \leq \theta \leq 0} |\phi(\theta)|$$

where $|\phi(\theta)| = \sum_{i=1}^{n} |\phi_i(\theta)|$ is a normal in $R^n$.

The initial condition for (6.41) and (6.42) is assumed to be

$$u_i(t) = \phi_i(t), -\tau \leq t \leq 0, i = 1, \cdots, n$$

where $\phi = (\phi_1, \cdots, \phi_n)^T \in C$. For any initial value $\phi \in C$, denote the solution of (6.41) or (6.42) starting from $\phi$ by $U(t, \phi) = (u_1(t, \phi), \cdots, u_n(t, \phi))^T$. Define

$$U_t(\phi) = U(t + \theta, \phi), \quad \theta \in [-\tau, 0]$$

for $t \geq 0$. Then, $U_t(\phi) \in C$ for each $t \geq 0$.

A point $U^* = (u_1^*, \cdots, u_n^*)^T$ is an equilibrium of (6.42), if it satisfies

$$\begin{cases} -d_i u_i^* + \sum_{j=1}^{n} (a_{ij} + b_{ij})\, v_j^* + I_i \equiv 0 \\ v_i^* = g_i(u_i^*), (i = 1, \cdots, n). \end{cases}$$

Obviously, $U^* \in C$. It is easy to see that $U^*$ is an equilibrium of (6.42) if and only if the solution of (6.42) starting from $U^*$ satisfies $U(t, U^*) \equiv U^*$ for all $t \geq 0$, i.e., $U(t, U^*)$ is a constant solution of (6.42). It should be noted here that the network (6.41) does not have any equilibrium point.

DEFINITION 6.7 *The network (6.41) is said to be absolutely periodic, if for every $g_i \in S(i = 1, \cdots, n)$, every $I(t) \in P$, every $d_i > 0 (i = 1, \cdots, n)$, and every $\tau_{ij} \geq 0(i, j = 1, \cdots, n)$, a unique periodic solution of network (6.41) exists and all other solutions of (6.41) converge to it exponentially as $t \to +\infty$.*

DEFINITION 6.8 *The network (6.42) is said to be absolutely stable, if for every $g_i \in S(i = 1, \cdots, n)$, every $I \in R^n$, every $d_i > 0(i = 1, \cdots, n)$, and every $\tau_{ij} \geq 0(i, j = 1, \cdots, n)$, a unique equilibrium of network (6.42) exists and this equilibrium is globally exponentially stable.*

Let $\Omega = (\omega_{ij})_{n \times n}$ be a matrix with the elements satisfy $\omega_{ii} \geq 0(i = 1, \cdots, n)$ and $\omega_{ij} \leq 0(i \neq j; i, j = 1, \cdots, n)$. $\Omega$ is called an *M-matrix* [20], if the real part of each eigenvalue of $\Omega$ is nonnegative.

The following Lemma 6.6 and Lemma 6.7 about *M-matrix* are adapted from [20].

LEMMA 6.6 *The following conditions are equivalent*
*(I) $\Omega$ is a nonsingular M-matrix.*
*(II) Constants $\gamma_i > 0(i = 1, \cdots, n)$ exist, such that*

$$\gamma_i \omega_{ii} + \sum_{j=1, j \neq i}^{n} \gamma_j \omega_{ij} > 0, (i = 1, \cdots, n).$$

*(III) Constants $\gamma_i > 0(i = 1, \cdots, n)$ exist, such that*

$$\gamma_j \omega_{jj} + \sum_{i=1, i \neq j}^{n} \gamma_i \omega_{ij} > 0, (j = 1, \cdots, n).$$

LEMMA 6.7 *$\Omega$ is an M-matrix if and only if $\sigma E + \Omega$ is a nonsingular M-matrix for each $\sigma > 0$.*

LEMMA 6.8 *If constants $\gamma_i > 0(i = 1, \cdots, n)$ exist, such that*

$$\gamma_i \omega_{ii} + \sum_{j=1, j \neq i}^{n} \gamma_j \omega_{ij} \geq 0, (i = 1, \cdots, n) \tag{6.43}$$

*or*

$$\gamma_j \omega_{jj} + \sum_{i=1, i \neq j}^{n} \gamma_i \omega_{ij} \geq 0, (j = 1, \cdots, n) \tag{6.44}$$

*then, $\Omega$ is an M-matrix.*

*Proof*: For any $\sigma > 0$, it is easy to see that

$$\gamma_i\left(\sigma + \omega_{ii}\right) + \sum_{j=1, j\neq i}^{n} \gamma_j \omega_{ij} > 0, (i = 1, \cdots, n)$$

or

$$\gamma_j\left(\sigma + \omega_{jj}\right) + \sum_{i=1, i\neq j}^{n} \gamma_i \omega_{ij} > 0, (j = 1, \cdots, n).$$

Then, by the Lemma 6.6, the matrix $\sigma E + \Omega$ must be a nonsingular *M-matrix*. Since $\sigma > 0$ is arbitrary, then by the Lemma6.7, $\Omega$ is an *M-matrix*. This completes the proof.

However, the inverse of Lemma 6.8 is not correct. For example, the matrix

$$\begin{bmatrix} 1 & 0 & -2 & 0 & 0 & 0 \\ 0 & 1 & -1 & 0 & 0 & 0 \\ 0 & -1 & 1 & 0 & 0 & 0 \\ 0 & 0 & 0 & 1 & 0 & 0 \\ 0 & 0 & 0 & 0 & 1 & -1 \\ 0 & 0 & 0 & -2 & -1 & 1 \end{bmatrix}$$

is an *M-matrix*, but it does not satisfy (6.43) or (6.44). Moreover, in the Lemma 6.8, inequality (6.43) is not equivalent to inequality (6.44). For example, the matrix

$$\begin{bmatrix} 1 & 0 & -2 \\ 0 & 1 & -1 \\ 0 & -1 & 1 \end{bmatrix}$$

is also an *M-matrix*, it satisfies inequality (6.43) but it does not satisfy inequality (6.44).

## 5.2 Main Results

Given any $\phi, \psi \in C$, let $U(t,\phi) = (u_1(t,\phi), \cdots, u_n(t,\phi))$ and $U(t,\psi) = (u_1(t,\psi), \cdots, u_n(t,\psi))$ be the solutions of (6.41) starting from $\phi$ and $\psi$, respectively. Denote

$$\begin{cases} \bar{u}_i(t) = u_i(t,\phi) - u_i(t,\psi) \\ \bar{v}_i(t) = v_i(t,\phi) - v_i(t,\psi), (i = 1, \cdots, n) \end{cases}$$

for all $t \geq 0$. Then, it follows from (6.41) that

$$\frac{d\bar{u}_i(t)}{dt} = -d_i \bar{u}_i(t) + \sum_{j=1}^{n} \left[a_{ij}\bar{v}_j(t) + b_{ij}\bar{v}_j(t - \tau_{ij})\right] \tag{6.45}$$

for $t \geq 0$ and $i = 1, \cdots, n$. From (6.45), it follows that

$$\begin{aligned} D^+|\bar{u}_i(t)| &\leq -d|\bar{u}_i(t)| + a_{ii}|\bar{v}_i(t)| \\ &+ \sum_{j=1, j\neq i}^{n} |a_{ij}||\bar{v}_j(t)| \\ &+ \sum_{j=1}^{n} |b_{ij}||\bar{v}_j(t-\tau_{ij})| \end{aligned} \tag{6.46}$$

for all $t \geq 0$, where $d = \min_{1 \leq i \leq n}(d_i) > 0$.

LEMMA 6.9 *Constructing the following Lyapunov functional*

$$V(t) = \sum_{i=1}^{n} \gamma_i \left[ |\bar{u}_i(t)|e^{\epsilon t} + e^{\epsilon\tau} \sum_{j=1}^{n} |b_{ij}| \int_{t-\tau_{ij}}^{t} |\bar{v}_j(s)|e^{\epsilon s} ds \right] \tag{6.47}$$

*for all $t \geq 0$, where $\epsilon > 0$ and $\gamma_i > 0 (i = 1, \cdots, n)$ are constants. If*

$$\frac{d-\epsilon}{k} - a_{jj} - \frac{1}{\gamma_j} \sum_{i=1}^{n} \gamma_i \Big( |a_{ij}|(1-\delta_{ij}) + e^{\epsilon\tau}|b_{ij}| \Big) > 0$$

*where $k > 0$ is a constant such that*

$$\bar{v}_i(t) \leq k\bar{u}_i(t), (i = 1, \cdots, n)$$

*for all $t \geq 0$, and*

$$\delta_{ij} = \begin{cases} 1, & i = j \\ 0, & i \neq j \end{cases}$$

*then, $V(t) \leq V(0)$ for all $t \geq 0$.*

*Proof*: From (6.46), it follows that

$$\begin{aligned} D^+\left(|\bar{u}_i(t)|e^{\epsilon t}\right) &\leq -\left(\frac{d-\epsilon}{k} - a_{ii}\right) |\bar{v}_i(t)|e^{\epsilon t} \\ &+ \sum_{j=1, j\neq i}^{n} |a_{ij}||\bar{v}_j(t)|e^{\epsilon t} \\ &+ e^{\epsilon\tau} \sum_{j=1}^{n} |b_{ij}||\bar{v}_j(t-\tau_{ij})|e^{\epsilon t} \end{aligned} \tag{6.48}$$

for all $t \geq 0$. Then, from (6.47) and (6.48) that

$$D^+V(t) \leq \sum_{i=1}^{n} \gamma_i \Bigg[ D^+\left(|\bar{u}_i(t)|e^{\epsilon t}\right)$$

$$
\begin{aligned}
& +e^{\epsilon t}\sum_{j=1}^{n}|b_{ij}|\Big(e^{\epsilon\tau}|\bar{v}_j(t)|-|\bar{v}_j(t-\tau_{ij})|\Big)\Bigg] \\
\leq\ & -e^{\epsilon t}\sum_{i=1}^{n}\gamma_i\Bigg[\left(\frac{d-\epsilon}{k}-a_{ii}\right)|\bar{v}_i(t)| \\
& -\sum_{j=1}^{n}\Big(|a_{ij}|(1-\delta_{ij})+e^{\epsilon\tau}|b_{ij}|\Big)|\bar{v}_j(t)|\Bigg] \\
\leq\ & -e^{\epsilon t}\sum_{j=1}^{n}\gamma_j|\bar{v}_j(t)|\Bigg[\frac{d-\epsilon}{k}-a_{jj} \\
& -\frac{1}{\gamma_j}\sum_{i=1}^{n}\gamma_i\Big(|a_{ij}|(1-\delta_{ij})+e^{\epsilon\tau}|b_{ij}|\Big)\Bigg]. \\
\leq\ & 0
\end{aligned}
$$

for all $t \geq 0$, and so $V(t) \leq V(0)$ for all $t \geq 0$. The proof is completed.

LEMMA 6.10 *Given any $\phi, \phi \in C$, if constants $\beta > 0$ and $\epsilon > 0$ exist, such that*

$$\| U_t(\phi) - U_t(\psi) \| \leq \beta e^{-\epsilon(t-\tau)} \| \phi - \psi \| \tag{6.49}$$

*for all $t \geq 0$, then a unique periodic solution of network (6.41) exists and all other solutions converge to it exponentially as $t \to +\infty$.*

*Proof*: Select a positive integer $m$ such that

$$\beta e^{-\epsilon(m\varpi-\tau)} \leq \frac{1}{2}.$$

Define a *Poincaré* mapping $T : C \to C$ by

$$T\phi = U_{\varpi}(\phi).$$

Then, from (6.49), it follows that

$$\| T^m\phi - T^m\psi \| \leq \frac{1}{2} \| \phi - \psi \| .$$

This shows that $T^m$ is a contraction mapping. Then, $T^m$ must have a unique fixed point in $C$, that is a unique $\phi^* \in C$ exists such that $T^m\phi^* = \phi^*$. Since

$$T^m(T\phi^*) = T(T^m\phi^*) = T\phi^*$$

then, $T\phi^* \in C$ is also a fixed point of $T^m$. By the uniqueness of the fixed point of the mapping $T^m$, we have $T\phi^* = \phi^*$. That is $U_{\varpi}(\phi^*) = \phi^*$. Let $U(t, \phi^*)$ be the solution of (6.41) starting from $\phi^*$. By using $I(t+\varpi) = I(t)$ for all $t \geq 0$, it follows that $U(t+\varpi,\phi^*)$ is also a solution of (6.41). Since $U_{t+\varpi}(\phi^*) = U_t(U_\varpi(\phi^*)) = U_t(\phi^*)$ for all $t \geq 0$, then it must have $U(t+\varpi,\phi^*) = U(t,\phi^*)$ for all $t \geq 0$. This shows that $U(t,\phi^*)$ is a periodic solution with period $\varpi$. It is easy to see from (6.49) that all other solutions converge to this periodic solution exponentially as $t \to +\infty$. The proof is completed.

Next, for the network (6.41) and (6.42), define a matrix $M = (m_{ij})_{n\times n}$, where

$$m_{ij} = \begin{cases} -a_{ii} - |b_{ii}|, & i = j \\ -|a_{ij}| - |b_{ij}|, & i \neq j. \end{cases}$$

THEOREM 6.11 *If $M$ is an M-matrix, then the network (6.41) is absolutely periodic.*

*Proof*: For every $g_i \in S(i = 1,\cdots,n)$, every $I(t) = (I_1(t),\cdots,I_n(t)) \in P$, every $d_i > 0(i = 1,\cdots,n)$, and every $\tau_{ij} \geq 0(i,j = 1,\cdots,n)$, consider the network (6.41).

Since $g_i \in S(i = 1,\cdots,n)$, then a constant $k > 0$ exists, such that $\bar{v}_i(t) \leq k\bar{u}_i(t), (i = 1,\cdots,n)$ for all $t \geq 0$.

Using the condition that $M$ is an *M-matrix*, then by Lemma 6.7, the matrix $(d/k)E + M$ is a nonsingular *M-matrix*, where $E$ is the $n \times n$ identity matrix. Using Lemma 6.6, constants $\gamma_i > 0(i = 1,\cdots,n)$ exist, such that

$$\gamma_j\left(\frac{d}{k} + m_{jj}\right) + \sum_{i=1,i\neq j}^{n} \gamma_i m_{ij} > 0, (j = 1,\cdots,n).$$

That is

$$\frac{d}{k} - a_{jj} - \frac{1}{\gamma_j}\sum_{i=1}^{n} \gamma_i\Big(|a_{ij}|(1-\delta_{ij}) + |b_{ij}|\Big) > 0$$

for $j = 1,\cdots,n$. It is easy to see that there must exist a $\epsilon > 0$ such that

$$\frac{d-\epsilon}{k} - a_{jj} - \frac{1}{\gamma_j}\sum_{i=1}^{n} \gamma_i\Big[|a_{ij}|(1-\delta_{ij}) + e^{\epsilon\tau}|b_{ij}|\Big] > 0$$

for $j = 1,\cdots,n$.

Using the above $\epsilon$, construct a Lyapunov functional

$$V(t) = \sum_{i=1}^{n} \gamma_i \left[|\bar{u}_i(t)|e^{\epsilon t} + e^{\epsilon\tau}\sum_{j=1}^{n}|b_{ij}|\int_{t-\tau_{ij}}^{t} |\bar{v}_j(s)|e^{\epsilon s}ds\right]$$

for all $t \geq 0$. By Lemma 6.9, $V(t) \leq V(0)$ for all $t \geq 0$.

Since

$$\begin{aligned} V(0) &\leq \sum_{i=1}^{n} \gamma_i \left[|\bar{u}_i(0)| + e^{\epsilon\tau}\sum_{j=1}^{n}|b_{ij}|\int_{-\tau}^{0} |\bar{v}_j(s)|e^{\epsilon s}ds\right] \\ &\leq \beta_1 \parallel \phi - \psi \parallel \end{aligned}$$

where

$$\beta_1 = \max_{1 \le i \le n}(\gamma_i) + \frac{ke^{\epsilon\tau}}{\epsilon}\sum_{i=1}^{n}\gamma_i \max_{1\le j\le n}(|b_{ij}|)$$

and

$$V(t) \ge e^{\epsilon t}\sum_{i=1}^{n}\gamma_i|\bar{u}_i(t)| \ge \gamma e^{\epsilon t}|U(t,\phi) - U(t,\psi)|$$

for all $t \ge 0$, where $\gamma = \min_{1\le i\le n}(\gamma_i)$, then

$$|U(t,\phi) - U(t,\psi)| \le \beta \parallel \phi - \psi \parallel e^{-\epsilon t} \tag{6.50}$$

for all $t \ge 0$, where $\beta = \beta_1/\gamma$.

From (6.50), it follows that

$$\begin{aligned} \parallel U_t(\phi) - U_t(\psi) \parallel &= \sup_{-\tau\le\theta\le 0}|U(t+\theta,\phi) - U(t+\theta,\psi)| \\ &\le \beta e^{-\epsilon(t-\tau)} \parallel \phi - \psi \parallel \end{aligned} \tag{6.51}$$

for all $t \ge 0$. By (6.51), apply the Lemma 6.10, the result now follows and the proof is completed.

THEOREM 6.12 *If $M$ is an M-matrix, then the network (6.42) is absolutely stable.*

*Proof*: For every $g_i \in S(i = 1, \cdots, n)$, every $I = (I_1, \cdots, I_n) \in R^n$, every $d_i > 0(i = 1, \cdots, n)$, and every $\tau_{ij} \ge 0(i, j = 1, \cdots, n)$, consider the network (6.42). Since $(I_i, \cdots, I_n)^T$ is a constant vector, then for any constant $r > 0$, it holds that $I = I(t + r) = I(t)$ for all $t \ge 0$. Similarly to the proof of Theorem 6.11, it follows that a unique stable periodic solution $U(t, \phi^*)$ with period $r$ of network (6.42) exists. That is $U(t, \phi^*)$ is a periodic solution with any positive constants as its periods. Hence, it must be a constant solution, i.e, it is an equilibrium. Obviously, all other solutions globally exponentially converge to this equilibrium as $t \to +\infty$. This shows the network (6.42) is absolutely stable. The proof is completed.

THEOREM 6.13 *If constants $\gamma_i > 0(i = 1, \cdots, n)$ exist, such that*

$$\gamma_i m_{ii} + \sum_{j=1, j\ne i}^{n}\gamma_j m_{ij} \ge 0, (i = 1, \cdots, n)$$

*or*

$$\gamma_j m_{jj} + \sum_{i=1, i\ne j}^{n}\gamma_i m_{ij} \ge 0, (j = 1, \cdots, n).$$

*then the network (6.41) is absolutely periodic, and the network (6.42) is absolutely stable.*

*Proof*: By Lemma 6.8, the conditions imply that the matrix $M$ is an *M-matrix*. The result now follows directly from Theorem 6.11 and Theorem 6.12. The proof is completed.

Recently in [92], some conditions for absolute stability of delayed neural networks are reported. The absolute stability condition in [92] corresponding to the neural network (6.42) is that there exists a positive diagonal matrix $P$ such that

$$PA + AP^T + [PB]^T PB + E \tag{6.52}$$

is negative definite, where $A = (a_{ij})_{n \times n}$, $B = (b_{ij})_{n \times n}$ and $E$ is the $n \times n$ identity matrix.

The conditions in Theorem 6.13 for absolute stability are quite different from the above one. To make this clear, let's consider the simple one dimensional delayed neural network

$$\frac{du(t)}{dt} = -du(t) + ag[u(t)] + bg[u(t-\tau)] + I$$

where $d > 0$, $a$ and $b$ are constant. By Theorem 6.12, our condition for absolute stability is given by

$$a + |b| \le 0. \tag{6.53}$$

The condition (6.52) for this example becomes that there exists a constant $p > 0$ such that

$$2pa + p^2b^2 + 1 < 0. \tag{6.54}$$

If $a = -|b|$, our condition (6.53) holds, but the condition (6.54) will never hold for any $p > 0$.

Moreover, it is easy to see that our absolute stability conditions for delayed neural networks are consistent with the absolute stability condition for neural networks without delays that recently reported in [113].

## 5.3 Simulation Study on Absolute Periodicity

Some simulation results will be given in this subsection to further illustrate the absolute periodicity of the network (6.41). For the reason of simplicity, consider the simple three dimensional network as follows

$$\frac{du_i(t)}{dt} = -d_i u_i(t) + \sum_{j=1}^{3} a_{ij} g[u_j(t)] + I_i(t), (i = 1, 2, 3) \tag{6.55}$$

for $t \ge 0$.

The connection matrix $A = (a_{ij})_{3\times 3}$ of the neural network of (6.55) is randomly generated by the formula

$$\begin{cases} h_{ij} : \text{Normal}(0,1) \text{ for } i = 1,2 \text{ and } j = 1,2,3 \\ A = -Q\left[H_1 \times H_1^T + H_2 \times H_2^T\right] \end{cases}$$

where $H_i = (h_{i1}, h_{i2}, h_{i3})^T (i = 1,2)$ and

$$Q = \begin{pmatrix} 1 & 1 & 0 \\ 0 & 1 & 0 \\ 0 & 0 & 1 \end{pmatrix}.$$

It is not difficult to see that any matrix generated by this method is an *M-matrix*. Take an example

$$A = \begin{pmatrix} -1.5963 & -0.7327 & -1.3708 \\ -0.4619 & -0.2708 & -0.4341 \\ -0.9366 & -0.4341 & -0.8070 \end{pmatrix}$$

it's eigenvalues are $-2.6252$, $0$ and $-0.0488$. This matrix will be used for the following simulations.

In the network (6.55), the parameter $d_i (i = 1,2,3)$ will be randomly generated. Four types of activation functions $(e^s - e^{-s})/(e^s + e^{-s})$, $1/(1+exp(-s))$, $(|s+1| - |s-1|)/2$ and $\max(0, s)$ will be used to simulate the performance of the neural network (6.55). The input periodic vector functions will take elements as $\sin t$, $\cos t$ or their combinations.

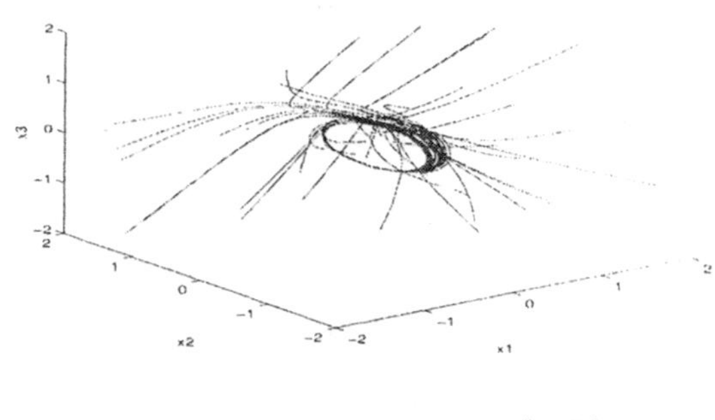

*Figure 6.3.* Network with $g = \frac{e^s - e^{-s}}{e^s + e^{-s}}$

Figure 6.3 and Figure 6.4 are the simulation results with

$$\begin{cases} g(s) = \frac{e^s - e^{-s}}{e^s + e^{-s}} \\ I(t) = (\sin t, \cos t, \sin t)^T \\ d_1 = 0.9299, d_2 = 0.7401, d_3 = 0.9588 \end{cases}$$

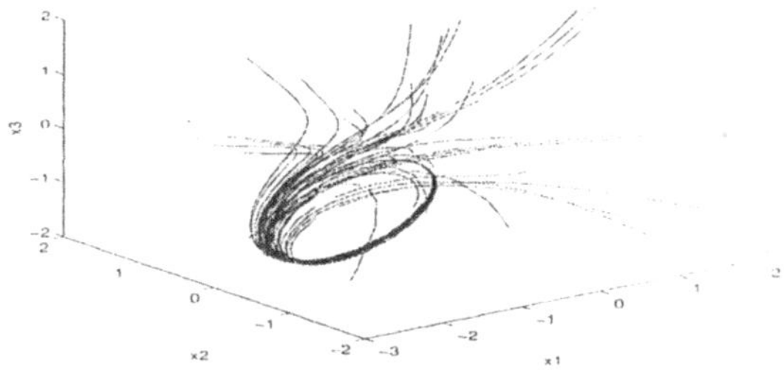

*Figure 6.4.* Network with $g = \frac{1}{1+e^{-s}}$

and

$$\begin{cases} g(s) = \frac{1}{1+e^{-s}} \\ I(t) = (\cos t, \sin t, \cos t)^T \\ d_1 = 0.6419, d_2 = 0.9599, d_3 = 0.7325 \end{cases}$$

respectively. The activation functions are sigmoid function, which are continuous, bounded, differentiable and increasing functions. These types of activation functions are frequently used in Hopfield RNNs. The existence of stable periodic solutions can be clearly observed in Figure 6.3 and Figure 6.4.

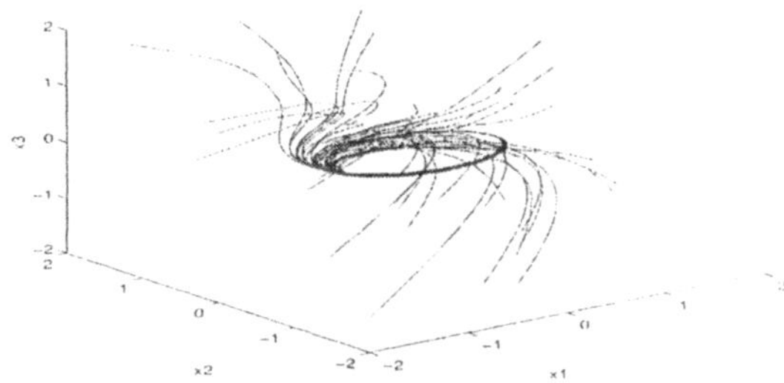

*Figure 6.5.* Network with $g = \frac{|s+1|-|s-1|}{2}$

Figure 6.5 is the simulation result of (6.55) with

$$\begin{cases} g(s) = \frac{|s+1|-|s-1|}{2} \\ I(t) = (\cos t, \sin t, \cos t)^T \\ d_1 = 0.9538, d_2 = 0.6807, d_3 = 0.9310 \end{cases}$$

where $g(s)$ is a bounded continuous nondecreasing function, which is often used in the CNNs. Also, the existence of a stable periodic solution can be clearly seen.

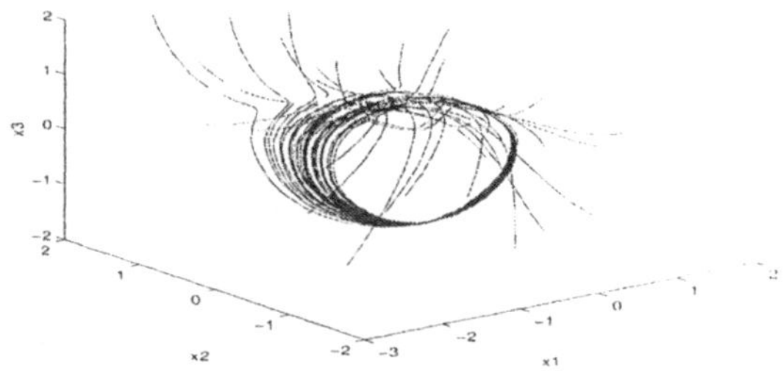

*Figure 6.6.* Network with $g = \max(0, s)$

Figure 6.6 is the simulation result of (6.55) with

$$\begin{cases} g(s) = \max(0, s) \\ I(t) = (\cos t, \sin t, \sin t + \cos t)^T \\ d_1 = 0.6847, d_2 = 0.8682, d_3 = 0.9820 \end{cases}$$

where the activation function is an unbounded nondecreasing function. It can be observed in Figure 6.6 that a stable periodic solution of the network (6.55) exists.

Besides the above simulation results, the network (6.55) has been widely simulated by using many other types of activation functions in $S_{GLAF}$ and many other input periodic vector functions in $P$, the stable periodic solutions can be observed in all of our simulation results. This further confirmed that the network processes the property of absolute periodicity.

## 5.4 Conclusions

In this section, the absolute periodicity and absolute stability of RNNs with delays have been studied. Some simple and checkable conditions on networks' connection matrix to guarantee the absolute periodicity and absolute stability have been obtained. The fact that RNNs have the property of absolute periodicity is an interesting dynamic behavior and it is believed that this property will enable more applications in practice.

# 6. Bidirectional Associative Memory RNNs with Delays

This section discusses the convergence of bidirectional associative memory RNNs with delays. The RNN model is described by

$$\begin{cases} \dot{x}_i(t) = -a_i x_i(t) + \sum_{j=1}^{q} \omega_{ij} s_j(y_j(t - \tau_{ij}^{(1)}(t))) + I_i(t), (i = 1, \cdots, p) \\ \\ \dot{y}_j(t) = -b_j y_j(t) + \sum_{i=1}^{p} \nu_{ji} s_i(x_i(t - \tau_{ji}^{(2)}(t))) + J_j(t), (j = 1, \cdots, q) \end{cases} \tag{6.56}$$

for $t \geq 0$, where $x = (x_1, \cdots, x_p)^T \in R^p$, $y = (y_1, \cdots, y_q)^T \in R^q$, $a_i > 0$ and $b_i > 0$ are constant, the activation functions $s_i \in S_{GLAF}(i = 1, \cdots, p+q)$, $I_i(t)$ and $J_j(t)$ are external inputs, $\omega_{ij}$ and $\nu_{ij}$ are connection weights of the network, the network delays $\tau_{ij}^{(1)}(t)$ and $\tau_{ji}^{(2)}(t)$ are continuous functions which satisfy $0 \leq \tau_{ij}^{(1)}(t) \leq \tau$ and $0 \leq \tau_{ji}^{(2)}(t) \leq \tau$, where $\tau \geq 0$ is a constant.

The following condition will be used to through this section:

$$\frac{1}{a_i} \sum_{j=1}^{q} \left[ l_j |\omega_{ij}| \left( \frac{1}{b_j} \sum_{r=1}^{p} l_r |\nu_{rj}| \right) \right] < 1, (i = 1, \cdots, p). \tag{6.57}$$

## 6.1 Equilibrium Points and Stability Analysis

In this subsection, assume that $I_i(t) \equiv I_i$ and $J_j(t) \equiv J_j$ for all $t \geq 0$.

A point $(x_1^*, \cdots, x_p^*, y_1^*, \cdots, y_q^*)^T \in R^{p+q}$ is an equilibrium of (6.56), if

$$\begin{cases} -a_i x_i^* + \sum_{j=1}^{q} \omega_{ij} s_j(y_j^*) + I_i = 0, (i = 1, \cdots, p) \\ \\ -b_j y_j^* + \sum_{i=1}^{p} \nu_{ji} s_i(x_i^*) + J_i = 0, (j = 1, \cdots, q). \end{cases}$$

Given any $t_0 \geq 0$, assume the initial condition of (6.56) as follows:

$$\begin{cases} x_i(t) = \alpha_i(t), (i = 1, \cdots, p) \\ \\ y_j(t) = \beta_j(t), (j = 1, \cdots, q) \end{cases}$$

for $t_0 - \tau \leq t \leq t_0$.

Denote

$$\Pi = \max_{1 \leq i \leq p; 1 \leq j \leq p} \left[ \sup_{t_0 - \tau \leq t \leq t_0} |\alpha_i(t) - x_i^*|, \sup_{t_0 - \tau \leq t \leq t_0} |\beta_j(t) - y_j^*| \right].$$

The network (6.56) is called global exponential stable, if for any $t_0 \geq 0$, the network (6.56) has only one equilibrium $(x_1^*, \cdots, x_p^*, y_1^*, \cdots, y_q^*)^T \in R^{p+q}$, and there exists constants $\epsilon > 0$ and $K \geq 1$ such that

$$\begin{cases} |x_i(t) - x_i^*| \leq K\Pi e^{-\epsilon(t-t_0)}, (i = 1, \cdots, p) \\ \\ |y_j(t) - y_j^*| \leq K\Pi e^{-\epsilon(t-t_0)}, (j = 1, \cdots, q) \end{cases}$$

for all $t \geq t_0$.

THEOREM 6.14 *If the condition (6.57) holds, then the network (6.56) has only one equilibrium.*

*Proof*: Construct the sequence

$$\left\{\left(x_1^{(k)}, \cdots, x_p^{(k)}, y_1^{(k)}, \cdots, y_q^{(k)}\right)\right\}, (k = 0, 1, 2, \cdots,)$$

by $x_i^{(0)} = 0 (i = 1, \cdots, p)$, $y_j^{(0)} = 0 (j = 1, \cdots, q)$, and

$$\begin{cases} x_i^{(k+1)} = \dfrac{1}{a_i} \displaystyle\sum_{j=1}^{q} \omega_{ij} s_j(y_j^{(k+1)}) + \dfrac{I_i}{a_i} \\ \\ y_j^{(k+1)} = \dfrac{1}{b_j} \displaystyle\sum_{i=1}^{p} \nu_{ji} s_i(x_i^{(k+1)}) + \dfrac{J_i}{b_j}. \end{cases} \tag{6.58}$$

Define series as

$$\sum_{k=0}^{\infty} \left[x_i^{(k+1)} - x_i^{(k)}\right], (i = 1, \cdots, p). \tag{6.59}$$

Next, it will be proven that each series is convergent.
By (6.58), it follows that

$$\begin{aligned} |x_i^{(k+1)} - x_i^{(k)}| &\leq \frac{1}{a_i} \sum_{j=1}^{q} l_j |\omega_{ij}| |y_j^{(k+1)} - y_j^{(k)}| \\ &\leq \frac{1}{a_i} \sum_{j=1}^{q} l_j |\omega_{ij}| \frac{1}{b_j} \sum_{r=1}^{p} l_r |\nu_{rj}| |x_r^{(k)} - x_r^{(k-1)}| \\ &\leq \frac{1}{a_i} \sum_{j=1}^{q} l_j |\omega_{ij}| \frac{1}{b_j} \sum_{r=1}^{p} l_r |\nu_{rj}| \|x^{(k)} - x^{(k-1)}\| \end{aligned}$$

where

$$\|x^{(k)} - x^{(k-1)}\| = \max_{1 \leq r \leq q} |x_r^{(k)} - x_r^{(k-1)}|.$$

Denote,

$$\delta = \max_{1\leq i\leq q}\left[\frac{1}{a_i}\sum_{j=1}^{q} l_j|\omega_{ij}|\frac{1}{b_j}\sum_{r=1}^{p} l_r|\nu_{rj}|\right] < 1.$$

Then,

$$\|x^{(k+1)} - x^{(k)}\| \leq \delta\|x^{(k)} - x^{(k-1)}\|.$$

Using mathematical induction, it follows that

$$|x_i^{(k+1)} - x_i^{(k)}| \leq \|x^{(k+1)} - x^{(k)}\| \leq \delta^k\|x^{(1)} - x^{(0)}\|, (i = 1, \cdots, p).$$

This shows that each series (6.59) is convergent. Since

$$x_i^{(k)} = \sum_{r=0}^{k}\left[x_i^{(r)} - x_i^{(r-1)}\right], (i = 1, \cdots, p),$$

then, $\left\{x_i^{(k)}\right\}(i = 1, \cdots, p)$ must converge. Suppose that $x_i^{(k)} \to x_i^*(i = 1, \cdots, p)$, then by (6.58), $\left\{y_j^{(k)}\right\}(j = 1, \cdots, q)$ must converge also. Suppose that $y_j^{(k)} \to y_j^*(j = 1, \cdots, q)$. Letting $k \to +\infty$, then by (6.58), it follows that

$$\begin{cases} -a_i x_i^* + \sum_{j=1}^{q}\omega_{ij}s_j(y_j^*) + I_i = 0, (i = 1, \cdots, p) \\ \\ -b_j y_j^* + \sum_{i=1}^{p}\nu_{ji}s_i(x_i^*) + J_j = 0, (j = 1, \cdots, q). \end{cases} \tag{6.60}$$

Thus, $(x_1^*, \cdots, x_p^*, y_1^*, \cdots, y_q^*)^T$ is an equilibrium of (6.56).

Suppose the network (6.56) has another equilibrium $(x_1^\dagger, \cdots, x_p^\dagger, y_1^\dagger, \cdots, y_q^\dagger)^T$, i.e.,

$$\begin{cases} -a_i x_i^\dagger + \sum_{j=1}^{q}\omega_{ij}s_j(y_j^\dagger) + I_i = 0, (i = 1, \cdots, p) \\ \\ -b_j y_j^\dagger + \sum_{i=1}^{p}\nu_{ji}s_i(x_i^\dagger) + J_i = 0, (j = 1, \cdots, q). \end{cases} \tag{6.61}$$

By (6.60) and (6.61), it follows that

$$\begin{aligned} |x_i^* - x_i^\dagger| &\leq \frac{1}{a_i}\sum_{j=1}^{q} l_j|\omega_{ij}||y_j^* - y_j^\dagger| \\ &\leq \frac{1}{a_i}\sum_{j=1}^{q}\left(l_j|\omega_{ij}|\frac{1}{b_j}\sum_{r=1}^{p} l_r|\nu_{rj}|\right)\|x^* - x^\dagger\| \\ &\leq \delta\|x^* - x^\dagger\| \end{aligned}$$

where $\|x^* - x^\dagger\| = \max_{1\leq r\leq q} |x_r^* - x_r^\dagger|$. Thus,

$$\|x^* - x^\dagger\| \leq \delta\|x^* - x^\dagger\|.$$

Since $\delta < 1$, it must be $\|x^* - x^\dagger\| = 0$, i.e., $x_i^* = x_i^\dagger (i = 1, \cdots, p)$. Using (6.60) and (6.61), it follows that $y_j^* = y_j^\dagger (j = 1, \cdots, q)$. The proof is completed.

THEOREM 6.15 *If the condition (6.57) holds, then the network (6.56) is globally exponentially stable.*

*Proof*: By condition (6.57), there must exit a constant $0 < \epsilon < a_i (i = 1, \cdots, p)$ such that

$$\delta(\epsilon) \triangleq \max_{1\leq i\leq p}\left[\frac{e^{\epsilon\tau}}{a_i - \epsilon}\sum_{j=1}^{p}\left(l_j|\omega_{ij}|\frac{e^{\epsilon\tau}}{b_j - \epsilon}\sum_{r=1}^{q} l_r|\nu_{jr}|\right)\right] < 1.$$

By Theorem 6.14, the network (6.56) has a unique equilibrium

$$(x_1^*, \cdots, x_p^*, y_1^*, \cdots, y_q^*)^T \in R^{p+q}.$$

Given any $t_0 \geq 0$, define

$$\begin{cases} P_i(t) = |x_i(t) - x_i^*|e^{\epsilon(t-t_0)} \\ Q_j(t) = |y_j(t) - y_j^*|e^{\epsilon(t-t_0)} \end{cases}$$

for $t \geq t_0 - \tau$. Then, it follows that

$$D^+P_i(t) \leq -(a_i - \epsilon)P_i(t) + e^{\epsilon\tau}\sum_{j=1}^{q} l_j|\omega_{ij}|Q_j(t - \tau_{ij}^{(1)}(t)), (i = 1, \cdots, p) \tag{6.62}$$

and

$$D^+Q_j(t) \leq -(b_j - \epsilon)Q_j(t) + e^{\epsilon\tau}\sum_{i=1}^{p} l_i|\nu_{ji}|P_j(t - \tau_{ji}^{(2)}(t)), (j = 1, \cdots, q) \tag{6.63}$$

for all $t \geq t_0$. Let

$$\begin{cases} P(t) = \max_{1\leq i\leq p}\left[\sup_{t_0-\tau\leq s\leq t}(P_i(s))\right] \\ Q(t) = \max_{1\leq j\leq q}\left[\sup_{t_0-\tau\leq s\leq t}(Q_i(s))\right] \end{cases}$$

for $t \geq t_0$. Clearly, $P(t)$ and $Q(t)$ are monotone non-decrease. By (6.62), it follows that

$$\begin{aligned} P_i(t) &\leq e^{-(a_i-\epsilon)}P_i(t_0) + e^{\epsilon\tau}\sum_{j=1}^{q} l_j|\omega_{ij}| \\ &\quad \times \int_{t_0}^{t} e^{-(a_i-\epsilon)(t-s)}Q_j(s-\tau_{ij}^{(1)}(s))ds \\ &\leq \Pi + \frac{e^{\epsilon\tau}}{a_i-\epsilon}\sum_{j=1}^{q} l_j|\omega_{ij}| \sup_{t_0-\tau\leq s\leq t}(Q_j(s)), \end{aligned} \tag{6.64}$$

for $t \geq t_0$ and $(i = 1, \cdots, p)$. By (6.63), it follows that

$$\begin{aligned} Q_j(t) &\leq \Pi + \frac{e^{\epsilon\tau}}{b_j-\epsilon}\sum_{i=1}^{p} l_i|\nu_{ji}| \sup_{t_0-\tau\leq s\leq t}(P_i(s)) \\ &\leq \Pi + \frac{e^{\epsilon\tau}}{b_j-\epsilon}\sum_{i=1}^{p} l_i|\nu_{ji}|P(t), (i = 1, \cdots, p) \end{aligned} \tag{6.65}$$

for $t \geq t_0$. Noting that $Q_j(t) \leq \Pi (j = 1, \cdots, p)$ for $t_0 - \tau \leq t \leq t_0$, substitute (6.65) into (6.64), it follows that

$$\begin{aligned} P_i(t) &\leq \left(1 + \frac{e^{\epsilon\tau}}{a_i-\epsilon}\sum_{j=1}^{q} l_j|\omega_{ij}|\right)\Pi \\ &\quad + \frac{e^{\epsilon\tau}}{a_i-\epsilon}\sum_{j=1}^{p}\left(l_j|\omega_{ij}|\frac{e^{\epsilon\tau}}{b_j-\epsilon}\sum_{r=1}^{q} l_r|\nu_{jr}|\right) \\ &\leq \left(1 + \frac{e^{\epsilon\tau}}{a_i-\epsilon}\sum_{j=1}^{q} l_j|\omega_{ij}|\right)\Pi + \delta(\epsilon)P(t) \end{aligned}$$

for $t \geq t_0$. Since $P_i(t) \leq \Pi (i = 1, \cdots, p)$, then

$$P(t) \leq \max_{1\leq i\leq p}\left(1 + \frac{e^{\epsilon\tau}}{a_i-\epsilon}\sum_{j=1}^{q} l_j|\omega_{ij}|\right)\Pi + \delta(\epsilon)P(t)$$

for $t \geq t_0$, i.e.,

$$P(t) \leq \frac{1}{1-\delta(\epsilon)}\max_{1\leq i\leq p}\left(1 + \frac{e^{\epsilon\tau}}{a_i-\epsilon}\sum_{j=1}^{q} l_j|\omega_{ij}|\right)\Pi$$

for $t \geq t_0$. Thus, it follows that

$$|x_i(t) - x_i^*| \leq \frac{\Pi}{1-\delta(\epsilon)}\max_{1\leq i\leq p}\left(1 + \frac{e^{\epsilon\tau}}{a_i-\epsilon}\sum_{j=1}^{q} l_j|\omega_{ij}|\right)e^{-\epsilon(t-t_0)}, (i = 1, \cdots, p)$$

for $t \geq t_0$. Similarly, it is easy to show that

$$|y_j(t) - y_j^*| \leq K\Pi e^{-\epsilon(t-t_0)}, (j = 1, \cdots, q)$$

for all $t \geq t_0$, where $K \geq 1$ is some constant. Thus, the network (6.56) is globally exponentially stable. The proof is completed.

## 6.2 Stable Periodic Trajectories

This subsection considers the existence of stable periodic trajectories of the network (6.56) under the assumption that $I_i(t + \varpi) \equiv I_i(t)$ and $J_j(t + \varpi) \equiv J_j(t)$, where $\varpi > 0$ is a constant.

The network (6.56) is said to have a stable periodic trajectory, if (6.56) has a periodic trajectory and all other trajectories converge to it.

THEOREM 6.16 *If the condition (6.57) holds, then the network (6.56) has a stable periodic trajectory.*

*Proof*: Let $C = C([-\tau, 0], R^{p+q})$ be the *Banach* space of continuous functions mapping the interval $[-\tau, 0]$ into $R^{p+q}$ with the topology of uniform convergence. Given any $\phi \in C$, denote

$$\|\phi\| = \sup_{-\tau \leq \theta \leq 0} (|\phi(\theta)|)$$

where $|\phi(\theta)| = \sum_{i=1}^{p+q} |\phi_i(\theta)|$. For any $\phi, \psi \in C$, denote

$$U(t, \phi) = (x_1(t, \phi), \cdots, x_p(t, \phi), y_1(t, \phi), \cdots, y_q(t, \phi))^T$$

and

$$U(t, \psi) = (x_1(t, \psi), \cdots, x_p(t, \psi), y_1(t, \psi), \cdots, y_q(t, \psi))^T$$

be the trajectories of (6.56) passing through $(0, \phi)$ and $(0, \psi)$, respectively. Define $U_t(\phi) = U(t + \theta, \phi), \theta \in [-\tau, 0], t \geq 0$, clearly, $U_t(\phi) \in C$.

From (6.56), it follows that

$$\begin{aligned} & \frac{d\left[x_i(t, \phi) - x_i(t, \psi)\right]}{dt} \\ = & -a_i\left[x_i(t, \phi) - x_i(t, \psi)\right] \\ & + \sum_{j=1}^{q} w_{ij}\left[s_j(y_j(t - \tau_{ij}^{(1)}, \phi)) - s_j(y_j(t - \tau_{ij}^{(1)}, \psi)\right] \end{aligned}$$

and

$$\begin{aligned} & \frac{d\left[y_j(t, \phi) - y_j(t, \psi)\right]}{dt} \\ = & -b_j\left[y_j(t, \phi) - y_j(t, \psi)\right] \\ & + \sum_{j=1}^{p} \nu_{ji}\left[s_j(x_i(t - \tau_{ji}^{(2)}, \phi)) - s_i(x_i(t - \tau_{ji}^{(2)}, \psi)\right]. \end{aligned}$$

Then, similarly to the proof of Theorem 6.15, there exist constants $\epsilon > 0$ and $K \geq 1$ such that

$$\|U_t(\phi) - U_t(\psi)\| \leq Ke^{-\epsilon t}\|\phi - \psi\| \tag{6.66}$$

for $t \geq 0$.

Taking a positive integer $m$ such that $Ke^{-\epsilon m\varpi} \leq 1/2$. Define *Poincare* mapping $T : C \to C$ by

$$T\phi = U_\varpi(\phi).$$

By (6.66), it follows that

$$\|T^m\phi - T^m\psi\| \leq \frac{1}{2}\|\phi - \psi\|.$$

Thus, $T^m$ is a contraction mapping. Then, there exists a unique $\phi^* \in C$ such that $T^m\phi^* = \phi^*$. Since $T^m(T\phi^*) = T(T^m\phi^*) = T\phi^*$, then $T\phi^* \in C$ is also a fixed point of $T^m$. Thus, $T\phi^* = \phi^*$, i.e, $U_\omega(\phi^*) = \phi^*$. Let $U(t, \phi^*)$ be the trajectory of (6.56) starting from $\phi^*$. Using $I(t+\varpi) = I(t)$, then, $U(t+\varpi, \phi^*)$ is also a solution of (6.56). Since $U_{t+\varpi}(\phi^*) = U_t(U_\varpi(\phi^*)) = U_t(\phi^*)$ for all $t \geq 0$, it follows that $U(t+\varpi, \phi^*) = U(t, \phi^*)$ for all $t \geq 0$. This shows that $U(t, \phi^*)$ is a periodic trajectory with period $\varpi$. Clearly, from (6.64), all other trajectories converge to this periodic trajectory exponentially as $t \to +\infty$. The proof is completed.

## 6.3 Discussions and Examples

All the results of this section are based on the condition (6.57). This condition relates both $\omega_{ij}$ and $\nu_{ji}$, thus, it can make bidirectional adjustment in among the connections of the network. Let $\gamma > 0$ be a constant, then the condition

$$\begin{cases} \dfrac{1}{a_i}\displaystyle\sum_{j=1}^{q} l_j|\omega_{ij}| < \dfrac{1}{\gamma}, (i = 1, \cdots, p) \\ \\ \dfrac{1}{b_j}\displaystyle\sum_{i=1}^{p} l_i|\nu_{ij}| \leq \gamma, (j = 1, \cdots, q) \end{cases}$$

implies that condition (6.57) holds. In particular, the condition

$$\begin{cases} \dfrac{1}{a_i}\displaystyle\sum_{j=1}^{q} l_j|\omega_{ij}| < 1, (i = 1, \cdots, p) \\ \\ \dfrac{1}{b_j}\displaystyle\sum_{i=1}^{p} l_i|\nu_{ij}| \leq 1, (j = 1, \cdots, q) \end{cases}$$

implies that condition (6.57) holds. This shows that condition (6.57) is weaker.

EXAMPLE 6.7 *Consider the network*

$$\begin{cases} \dot{x}(t) = -x(t) + 2s((y(t - \sin^2 t))) + 1 - \frac{\pi}{2} \\ \\ \dot{y}(t) = -y(t) + s((x(t - \cos^2 t))) + 3 - \frac{\pi}{4} \end{cases} \tag{6.67}$$

*for* $t \geq 0$, *where*

$$s(\theta) = \frac{e^{\theta} - e^{-\theta}}{e^{\theta} + e^{-\theta}}, s \in R.$$

Clearly, it holds that $|\dot{s}(\theta)| \leq 1$. Thus, condition (6.57) is satisfied. By Theorem 6.15, the network (6.67) is globally exponentially stable. It has $(1, 1)^T$ as its unique equilibrium.

EXAMPLE 6.8 *Consider the network*

$$\begin{cases} \dot{x}_1(t) = -5x_1(t) + s(y_1(t - \tau_1)) + s(y_2(t - \tau_2)) + \sin t \\ \dot{x}_2(t) = -7x_2(t) - s(y_1(t - \tau_3)) + s(y_2(t - \tau_4)) + \cos t \\ \dot{y}_1(t) = -y_1(t) + s(x_1(t - \tau_5)) - s(x_2(t - \tau_6)) + \sin t \\ \dot{y}_2(t) = -2y_2(t) + 3s(x_1(t - \tau_7)) + s(x_2(t - \tau_8)) + \cos t \end{cases} \tag{6.68}$$

*for* $t \geq 0$, *where* $\tau_i > 0 (i = 1, \cdots, 8)$ *are constants, and* $s$ *is defined as that in above example.*

It is not difficult to check that (6.68) satisfies the condition (6.57), thus, this network has a stable periodic trajectory.

Bidirectional associative memory RNNs are interesting RNNs. More stability results on bidirectional associative memory RNNs with delays can be found in [68, 173, 30].

Chapter 7

# OTHER MODELS OF CONTINUOUS TIME RECURRENT NEURAL NETWORKS

In this chapter, convergence of two other models of continuous time RNNs will be studied. In the first part, a model of RNNs with variable inputs and without decaying linear term is discussed. This model has been found applications in optimization problems. A concept, called input-to-state convergence (ISC), will be proposed and analyzed to this model. In the second part, a model which can be used to attract eigenvectors of any symmetric matrix is considered. A detailed mathematical analysis to the model will be provided.

## 1. RNNs with Variable Inputs

### 1.1 Introduction

The model of RNNs is described by the following nonlinear differential equation

$$\frac{dx_i(t)}{dt} = \sum_{j=1}^{n} a_{ij} g_j[x_j(t)] + I_i(t), (i = 1, \cdots, n) \tag{7.1}$$

for $t \geq 0$, where $a_{ij} (i, j = 1, \cdots, n)$ are the constant connection weights and $A = (a_{ij})_{n \times n}$ is the connection matrix, the inputs $I_i(t)(i = 1, \cdots, n)$ are some continuous functions defined on $[0, +\infty)$, and $g_i(i = 1, \cdots, n)$ are the network's activation functions.

This model of neural networks is generalized from the neural networks proposed in [192, 193, 194, 195, 196]. Special forms of (7.1) have been successfully used in [192, 193, 194, 195, 196] to solve various optimization problems. In [192, 193, 194, 195, 196], it has been demonstrated that this class of neural networks could be easily implemented via electronic circuits. This model is essentially different from the model of the well known Hopfield neural networks. In fact, since (7.1) contains time varying inputs, it is a non-autonomous differential equation. Moreover, it does not contain the linear terms as in Hopfield networks. Hence, the dynamic structure of this class of neural networks is different from that of Hopfield. Since different structures of differential equations usually result in different dynamic behavior, the convergence of (7.1) may be different from that of Hopfield.

It is well known that if the connection matrix of a Hopfield network is symmetric, then each trajectory will converge to a equilibrium. However, the network (7.1) does not possess this property. To illustrate this point, consider a simple one dimensional recurrent neural network with time varying input:

$$\frac{dx(t)}{dt} = -g[x(t)] - e^{-t},$$

for $t \geq 0$, where

$$g(x) = \frac{1}{1+e^{-x}}.$$

Obviously, the connection matrix is symmetric since the network is one dimensional. However,

$$\begin{aligned} \frac{d\left(e^{-x(t)} - x(t)\right)}{dt} &= -\left(1+e^{-x(t)}\right)\frac{dx(t)}{dt} \\ &\geq 1, \end{aligned}$$

for all $t \geq 0$. Then, $e^{-x(t)} - x(t) \to +\infty$ as $t \to +\infty$, i.e., $x(t) \to -\infty$ as $t \to +\infty$. This shows that any trajectory of this network will not converge.

In the case of Hopfield neural networks, equilibrium points always exist, and the trajectories are expected to converge to some of the equilibrium points. However, for the neural network (7.1), it does not possess any equilibrium points except when the time varying inputs become constants. An interesting problem to address for this network is that its convergence property.

Using convergence property of RNNs, neural networks can be used to solve optimization problems [25, 57, 60, 86, 102, 113, 114, 111, 136, 174, 184, 206, 81]. Therefore, convergence analysis is very important for neural networks. In practice, neural networks are often affected by noise, expressed for instance as perturbations on inputs and errors on observations. It is desired that no matter what the initial value is, if the inputs of a network converge, then the state of the network must converge. Recently, in the nonlinear control field, a new concept called input-to-state stability (ISS) was proposed [169]. Roughly, ISS means that no matter what the initial state is, if the inputs are uniformly small, then the state must eventually be small. ISS has been now widely recognized and accepted as an important concept that is useful for many nonlinear control problems. Many results on ISS have been reported, see for example [3, 168, 169, 170, 171]. Motivated by the ISS concept, in this chapter, a concept called input-to-state convergence (ISC) to (7.1) will be introduced in order to analyze its global convergence under the effect of time varying inputs. The problems to be addressed are under what conditions the trajectories of (7.1) will converge to a constant vector, and what kind of constant vector it would turn out to be.

## 1.2 Preliminaries

Let $A = (a_{ij})_{n\times n}$, $I(t) = (I_1(t), \cdots, I_n(t))^T$, $x(t) = (x_1(t), \cdots, x_n(t))^T$ and $G[x(t)] = (g_1[x_1(t)], \cdots, g_n[x_n(t)])^T$ then (7.1) can be rewritten in a compact vector form

$$\frac{dx(t)}{dt} = AG[x(t)] + I(t), \tag{7.2}$$

for $t \geq 0$.

Assume that each activation function $g_i$ is a sigmoid function defined by (2.2). Thus, there exists $g_i^{-1}$ and it is continuous.

For any $x(0) \in R^n$, $x(t, x(0))$ denotes the solution of (7.1) starting from $x(0)$.

DEFINITION 7.1 *The network (7.1) is said to be input-to-state converge (ISC), if*

$$\lim_{t\to+\infty} I(t) = I,$$

*and*

$$\Omega = \left\{y^* \in R^n \middle| AG(y^*) + I = 0\right\} \neq \phi,$$

*then, for any $x(0) \in R^n$, there exists a $x^* \in \Omega$ such that*

$$\lim_{t\to+\infty} x(t, x(0)) = x^*,$$

*for all $t \geq 0$.*

The ISC concept is similar to ISS [169] but tailored for the neural network (7.1). Since convergence is an interesting property for neural networks, ISC is a suitable concept to consider.

Denote

$$\begin{cases} \underline{g}_i = \inf\limits_{-\infty<s<+\infty} (g_i(s)) \\ \overline{g}_i = \sup\limits_{-\infty<s<+\infty} (g_i(s)), (i = 1, \cdots, n) \end{cases}$$

for all $s \in R$, where $\underline{g}_i$ and $\overline{g}_i$ may take the values $-\infty$ and $+\infty$, respectively.

The following *Lemma* gives a method to check that the set $\Omega$ is nonempty.

LEMMA 7.1 *Suppose $A^{-1} = (\tilde{a}_{ij})_{n\times n}$ exist, then $\Omega \neq \phi$ if and only if*

$$-\sum_{j=1}^{n} \tilde{a}_{ij} I_j \in \left(\underline{g}_i, \overline{g}_i\right), (i = 1, \cdots, n).$$

*Proof*: For each $i(1 \leq i \leq n)$, since

$$\underline{g}_i < -\sum_{j=1}^{n} \tilde{a}_{ij} I_j < \overline{g}_i,$$

then by the continuity of $g_i$, there must exist a constant $y_i^*$ such that

$$g_i(y_i^*) = -\sum_{j=1}^{n} \tilde{a}_{ij} I_j.$$

Thus, $G(y^*) = -A^{-1}I$, i.e., $AG(y^*) + I = 0$. This completes the proof.

The following lemma which was proven in [220] will be used. It provides an interesting property of the activation functions.

LEMMA 7.2

$$\int_{\eta}^{\xi} [g_i(s) - g_i(\xi)]\, ds \geq \frac{1}{2k_i} [g_i(\xi) - g_i(\eta)]^2$$

*for all* $\xi, \eta \in R$ *and* $(i = 1, \cdots, n)$.

A matrix $M = (m_{ij})_{n \times n}$ is called an *M-matrix* if $m_{ii} \geq 0 (i = 1, \cdots, n)$, $m_{ij} \leq 0 (i \neq j; i, j = 1, \cdots, n)$, and the real part of each eigenvalue of $M$ is non-negative. $M$ is called a non-singular *M-matrix* if $M$ is both an *M-matrix* and non-singular. In [20], it gives that a matrix $M$ is a non-singular $M-matrix$ if and only if there exist constants $\alpha_i > 0 (i = 1, \cdots, n)$ such that

$$\alpha_i m_{ii} + \sum_{j=1, j \neq i}^{n} \alpha_j m_{ij} > 0, (i = 1, \cdots, n).$$

or

$$\alpha_j m_{jj} + \sum_{i=1, i \neq j}^{n} \alpha_i m_{ij} > 0, (j = 1, \cdots, n).$$

## 1.3 Convergence Analysis

In this section, some conditions will be derived to guarantee the ISC of the network (7.1).

THEOREM 7.1 *Let* $M = (m_{ij})_{n \times n}$, *where*

$$m_{ij} = \begin{cases} -a_{ii}, & \text{if } i = j \\ -|a_{ij}|, & \text{if } i \neq j. \end{cases}$$

*If* $M$ *is a non-singular M-matrix and* $\Omega \neq \phi$, *then the network (7.1) is ISC.*

*Proof*: Since $\Omega \neq \phi$, then a constant vector $y^* = (y_1^*, \cdots, y_n^*)^T \in R^n$ exists, such that

$$\sum_{j=1}^{n} a_{ij} g_i[y_j^*] + I_i = 0, (i = 1, \cdots, n).$$

For any $x(0) \in R^n$, let $x_i(t)(i = 1, \cdots, n)$ be the solution of (7.1) starting from $x(0)$. Then it follows from (7.1) that

$$\frac{d\left[x_i(t) - y_i^*\right]}{dt} = \sum_{j=1}^{n} a_{ij} \left[g_j(x_j(t)) - g_j(y_j^*)\right] + I_i(t) - I_i, \qquad (7.3)$$

for $t \geq 0$ and $(i = 1, \cdots, n)$.

Since $M$ is a nonsingular matrix, there exist constants $\alpha_i > 0$ such that

$$a_{ii} + \frac{1}{\alpha_i} \sum_{j \neq i; j=1}^{n} \alpha_j |a_{ij}| < 0.$$

Denote

$$\begin{cases} z_i(t) = \frac{1}{\alpha_i}[x_i(t) - y_i^*], \\ h_i(t) = \frac{1}{\alpha_i}[g_i[x_i(t)] - g_i[y_i^*]], \end{cases}$$

for $t \geq 0$ and $(i = 1, \cdots, n)$. From (7.3),

$$D^+|z_i(t)| \leq a_{ii}|h_i(t)| + \frac{1}{\alpha_i} \sum_{j=1, j\neq i}^{n} \alpha_j |a_{ij}||h_j(t)| + |I_i(t) - I_i|, \qquad (7.4)$$

and

$$\begin{aligned} D^+|h_i(t)| \ \leq \ & D^+ g_i(x_i(t)) \Bigg[ a_{ii}|h_i(t)| + |I_i(t) - I_i| \\ & + \frac{1}{\alpha_i} \sum_{j=1, j\neq i}^{n} \alpha_j |a_{ij}||h_j(t)| \Bigg], \end{aligned} \qquad (7.5)$$

for all $t \geq 0$ and $(i = 1, \cdots, n)$.

Since $I_i(t) \to I_i(i = 1, \cdots, n)$ as $t \to +\infty$, then a constant $m > 0$ exists, such that $|I_i(t) - I_i| \leq m(i = 1, \cdots, n)$ for all $t \geq 0$.

Denote

$$\delta_i = -\left[a_{ii} + \frac{1}{\alpha_i} \sum_{j=1, j\neq i} \alpha_j |a_{ij}|\right], (i = 1, \cdots, n)$$

then, $\delta_i > 0(i = 1, \cdots, n)$.

For any

$$D > \frac{m}{\min_{1 \le i \le n} (\delta_i)},$$

it will be shown that $|h_i(0)| < D(i = 1, \cdots, n)$ will imply $|h_i(t)| < D(i = 1, \cdots, n)$ for all $t \ge 0$. Otherwise, there must exist a $t_1 > 0$ such that $|h_i(t_1)| = D$, and

$$|h_j(t)| \begin{cases} < D, & 0 \le t < t_1, j = i \\ \le D, & 0 \le t \le t_1, i \ne j. \end{cases}$$

Thus, $D^+|h_i(t_1)| \ge 0$. However, from (7.5) it follows

$$\begin{aligned} D^+|h_i(t_1)| &\le D^+ g_i(x_i(t_1)) \cdot (-\delta_i D + m) \\ &< 0. \end{aligned}$$

This contradiction shows that $|h_i(t)|(i = 1, \cdots, n)$ are bounded for all $t \ge 0$.

Suppose that

$$\lim_{t \to +\infty} \sup |h_i(t)| = \sigma_i, (i = 1, \cdots, n).$$

Obviously, $0 \le \sigma_i < +\infty, (i = 1, \cdots, n)$.

Without loss of generality, assume that $\sigma_1 = \max_{1 \le i \le n}\{\sigma_i\}$. It will be proven that $\sigma_1 = 0$, then all $\sigma_i = 0(i = 2, \cdots, n)$. Otherwise, i.e., $\sigma_1 > 0$, it will be proven that it leads to a contradiction.

By $\lim_{t \to +\infty} I_i(t) = I_i(i = 1, \cdots, n)$, there must exist a $t_2 \ge 0$ such that

$$\frac{|I_i(t) - I_i|}{\delta_i} \le \frac{\sigma_1}{4}, (i = 1, \cdots, n),$$

for all $t \ge t_2$.

Choose a constant $\epsilon > 0$ such that

$$0 < \epsilon \le \frac{\alpha_i \delta_i \sigma_1}{4 \sum_{j=1}^{n} \alpha_j |a_{ij}|}, (i = 1, \cdots, n).$$

Since $\lim_{t \to +\infty} \sup |h_i(t)| = \sigma_i, (i = 1, \cdots, n)$, it is easy to see that there exists a $t_3 \ge t_2$, such that $t \ge t_3$ implies

$$|h_i(t)| \le \sigma_i + \epsilon \le \sigma_1 + \epsilon, 1 \le i \le n.$$

It may be claimed that there exists a $t_4 \ge t_3$ such that

$$D^+|h_1(t)| < 0,$$

for all $t \ge t_4$. Otherwise, there must exist a $t_5 > t_3$ such that

$$D^+|h_1(t_5)| \ge 0 \quad \text{and} \quad |h_1(t_5)| \ge \sigma_1 - \epsilon.$$

However, on the other hand, from (7.5), it follows

$$
\begin{aligned}
D^+|h_1(t_5)| &\le D^+g_1(x_1(t_5))\Bigg[a_{11}(\sigma_1-\epsilon) \\
&\quad +\frac{1}{\alpha_1}\sum_{j=1,j\neq 1}^{n}\alpha_j|a_{1j}|(\sigma_1+\epsilon)+\frac{\delta_1\sigma_1}{4}\Bigg] \\
&= D^+g_1(x_1(t_5))\Bigg[-\delta_1\sigma_1 \\
&\quad +\left(\frac{1}{\alpha_1}\sum_{j=1}^{n}\alpha_j|a_{1j}|\right)\epsilon+\frac{\delta_1\sigma_1}{4}\Bigg] \\
&\le -\frac{\delta_1\sigma_1}{2}\cdot D^+g_1(x_1(t_5)) \\
&< 0
\end{aligned}
$$

which is a contradiction. This shows that there exists a $t_4 \ge t_3$, such that $D^+|h_1(t)| < 0$ for all $t \ge t_4$. Thus,

$$
\lim_{t\to+\infty}|h_1(t)| = \lim_{t\to+\infty}\sup|h_1(t)| = \sigma_1.
$$

Obviously, there exists a $t_6 \ge t_4$ such that

$$
\begin{cases}
\sigma_1-\epsilon < |h_1(t)| < \sigma_1+\epsilon \\
|h_i(t)| < \sigma_1+\epsilon, (i=2,\cdots,n)
\end{cases}
$$

for all $t \ge t_6$.

From (7.4), it follows that

$$
\begin{aligned}
D^+|z_1(t)| &\le a_{11}(\sigma_1-\epsilon)+\frac{1}{\alpha_1}\sum_{j=1,j\neq 1}^{n}\alpha_j|a_{1j}|(\sigma_1+\epsilon) \\
&\quad +|I_1(t)-I_1| \\
&\le -\delta_1\sigma_1+\left(\frac{1}{\alpha_1}\sum_{j=1}^{n}\alpha_j|a_{1j}|\right)\epsilon+\frac{\delta_1\sigma_1}{4} \\
&\le -\frac{\delta_1\sigma_1}{2}
\end{aligned}
$$

for all $t \ge t_6$, and so

$$
\begin{aligned}
|z_1(t)| &\le |z_1(t_6)|-\frac{\delta_1\sigma_1}{2}(t-t_5) \\
&\to -\infty,
\end{aligned}
$$

as $t \to +\infty$. This contradicts $|z_1(t)| \geq 0$. Then, it must follow that $\sigma_1 = 0$, and so $\sigma_i = 0(i = 2, \cdots, n)$. This proves that

$$\lim_{t \to +\infty} |h_i(t)| = \lim_{t \to +\infty} \sup |h_i(t)| = 0, (i = 1, \cdots, n),$$

i.e.,

$$\lim_{t \to +\infty} g_i[x_i(t)] = g_i[y_i^*], (i = 1, \cdots, n).$$

Consequently,

$$\begin{aligned}
\lim_{t \to +\infty} x_i(t) &= \lim_{t \to +\infty} g_i^{-1}\left(g_i[x_i(t)]\right) \\
&= g_i^{-1}\left(\lim_{t \to +\infty} g_i[x_i(t)]\right) \\
&= g_i^{-1}\left(g_i[y_i^*]\right) \\
&= y_i^*.
\end{aligned}$$

This shows that the network (7.1) is ISC. The proof is completed.

THEOREM 7.2 *Suppose that*

$$\int_0^{+\infty} [I_i(t) - I_i]^2 \, ds < +\infty, (i = 1, \cdots, n).$$

*If the matrix $A + A^T$ is negative definite and $\Omega \neq \phi$, then, the network (7.1) is ISC.*

*Proof*: By $\Omega \neq \phi$, a constant vector $y^* \in \Omega$ exists such that

$$AG(y^*) + I = 0.$$

Rewrite (7.2) as

$$\frac{dx(t)}{dt} = A\left[G(x(t)) - G(y^*)\right] + I(t) - I, \tag{7.6}$$

where $G(y^*) = (g_1[y_1^*], \cdots, g_n[y_n^*])^T$.

Define a differentiable function

$$V(t) = \sum_{i=1}^{n} \int_{y_i^*}^{x_i(t)} [g_i(\theta) - g_i(y_i^*))] \, d\theta,$$

for $t \geq 0$. Obviously, $V(t) \geq 0$ for all $t \geq 0$.

Let $\lambda > 0$ be the minimal eigenvalue of the matrix $\left[A + A^T\right]/2$ and choose a constant $\beta > 0$ such that $\beta < \lambda$. It follows from (7.6) that

$$\begin{aligned}
\dot{V}(t) &= \sum_{i=1}^{n} \left(g_i[x_i(t)] - g_i[y_i^*]\right) \dot{x}_i(t) \\
&= [G(x(t)) - G(y^*)]^T \Big[A\,[G(x(t)) - G(y^*)] \\
&\quad + I(t) - I\Big] \\
&\leq [G(x(t)) - G(y^*)]^T \left[\frac{A + A^T}{2}\right] [G(x(t)) - G(y^*)] \\
&\quad + \|G(x(t)) - G(y^*)\| \|I(t) - I\| \\
&\leq -\lambda \|G(x(t)) - G(y^*)\|^2 \\
&\quad + \|G(x(t)) - G(y^*)\| \|I(t) - I\| \\
&\leq -(\lambda - \beta)\|G(x(t)) - G(y^*)\|^2 + \frac{\|I(t) - I\|^2}{4\beta},
\end{aligned}$$

for all $t \geq 0$, where $\|\cdot\|$ is a *Euclidean* norm.

Define another differentiable function

$$W(t) = \left[V(t) + \frac{1}{4\beta}\right] e^{-\int_0^t \|I(s)-I\|^2 ds},$$

for all $t \geq 0$. It is easy to see that

$$W(t) \geq \frac{1}{4\beta} e^{-\int_0^t \|I(s)-I\|^2 ds} > 0,$$

for $t \geq 0$.

Then, it follows

$$\begin{aligned}
\dot{W}(t) &\leq -\|I(t) - I\|^2 W(t) + \dot{V}(t) e^{-\int_0^t \|I(s)-I\|^2 ds} \\
&\leq -\frac{\|I(t) - I\|^2}{4\beta} e^{-\int_0^t \|I(s)-I\|^2 ds} \\
&\quad - \left[(\lambda - \beta)\|G(x(t)) - G(y^*)\|^2 - \frac{\|I(t) - I\|^2}{4\beta}\right] \\
&\quad \times e^{-\int_0^t \|I(s)-I\|^2 ds} \\
&\leq -(\lambda - \beta) e^{-\int_0^{+\infty} \|I(s)-I\|^2 ds} \|G(x(t)) - G(y^*)\|^2
\end{aligned}$$

for all $t \geq 0$.

By $\dot{W}(t) \leq 0$ for $t \geq 0$, it follows that $W(t) \leq W(0)$ for all $t \geq 0$. Then, it follows that

$$V(t) \leq \left[V(0) + \frac{1}{4\beta}\right] e^{\int_0^{+\infty} \|I(s)-I\|^2 ds},$$

for all $t \geq 0$. Using *Lemma 1*,

$$[g_i(x_i(t)) - g_i(y_i^*)]^2 \leq 2k_i \left[V(0) + \frac{1}{4\beta}\right] e^{\int_0^{+\infty} \|I(s)-I\|^2 ds},$$

for all $t \geq 0$ and $(i = 1, \cdots, n)$. This shows that $|g_i[x_i(t)] - g_i[y_i^*]|(i = 1, \cdots, n)$ are bounded. It is easy to see that $|\dot{x}_i(t)|(i = 1, \cdots, n)$ are bounded also.

It will be that $\|G(x(t)) - G(y^*)\|^2 \to 0$ as $t \to +\infty$. Suppose this is not true, then for some $\epsilon > 0$ there exists a divergent sequence $\{t_k\}$ for which $\|G(x(t_k) - G(y^*)\|^2 \geq 2\epsilon$. Since

$$\begin{aligned} & D^+\|G(x(t)) - G(y^*)\|^2 \\ \leq \; & 2\sum_{i=1}^{n} [g_i(x_i(t)) - g_i(y_i^*)] \cdot D^+ g_i(x_i(t)) \cdot \dot{x}_i(t), \end{aligned}$$

then, there exists a constant $\omega > 0$ such that

$$\left|D^+\|G(x(t)) - G(y^*)\|^2\right| \leq \omega.$$

Then, on the intervals

$$t_k \leq t \leq t_k + \frac{\epsilon}{\omega}, (k = 1, 2, \cdots),$$

it follows

$$\begin{aligned} \|G(x(t)) - G(y^*)\|^2 \; & \geq \; \|G(x(t_k)) - G(y^*)\|^2 \\ & \quad -\omega\,|t - t_k| \\ & \geq \; \epsilon. \end{aligned}$$

The above intervals can be assumed to be disjoint by using, if necessary, a subsequence of $t_k$.

Then, it follows that

$$W\left(t_k + \frac{\epsilon}{\omega}\right) - W(t_k) \leq -\frac{(\lambda - \beta)\epsilon^2}{\omega} e^{-\int_0^{+\infty} \|I(s)-I\|^2 ds} \tag{7.7}$$

for all $k = 1, 2, \cdots$.

Since $\dot{W}(t) \leq 0$ for all $t \geq 0$ and the intervals $[t_k, t_k + \frac{\epsilon}{\omega}](k = 1, 2, \cdots)$ are disjoint of each other, then

$$W(t_k) \leq W\left(t_{k-1} + \frac{\epsilon}{\omega}\right), \tag{7.8}$$

for all $k = 2, 3, \cdots$.

From (7.7) and (7.8), it follows that

$$\begin{aligned} & W\left(t_k + \frac{\epsilon}{2\omega}\right) - W(t_1) \\ \leq & -\frac{(\lambda - \beta)\epsilon^2}{\omega} e^{-\int_0^{+\infty} \|I(s) - I\|^2 ds} \cdot k \\ \rightarrow & -\infty \end{aligned}$$

as $k \rightarrow +\infty$, which contradicts to $W(t) > 0$. This proves that $G(x(t)) \rightarrow G(y^*)$ as $t \rightarrow +\infty$, i.e.,

$$\lim_{t \rightarrow +\infty} g_i(x_i(t)) = g_i(y_i^*), (i = 1, \cdots, n).$$

Thus,

$$\begin{aligned} \lim_{t \rightarrow +\infty} x_i(t) &= \lim_{t \rightarrow +\infty} g_i^{-1}(g_i[x_i(t)]) \\ &= g_i^{-1}\left(\lim_{t \rightarrow +\infty} g_i[x_i(t)]\right) \\ &= g_i^{-1}(g_i[y_i^*]) \\ &= y_i^*. \end{aligned}$$

This completes the proof.

It is easy to see that either the condition that $M$ is a non-singular *M-matrix* or the condition that $A + A^T$ is negative definite implies that $A^{-1}$ exist. Then, by *Lemma 7.2*, to check that $\Omega \neq \phi$ in Theorem 7.1 or Theorem 7.2, it is only necessary to check

$$-\sum_{j=1}^{n} \tilde{a}_{ij} b_j \in \left(\underline{g}_i, \overline{g}_i\right), (i = 1, \cdots, n)$$

where $\tilde{a}_{ij}(i, j = 1, \cdots, n)$ are elements of $A^{-1}$.

In Theorem 7.1, the condition added to the connection matrix is stronger than that in Theorem 7.2. However, the condition added to the inputs in Theorem 7.1 is weaker than that in Theorem 7.2.

## 1.4 Examples

In this subsection, some examples will be used to illustrate the theory in last subsection.

EXAMPLE 7.1 *Consider the following neural network with time varying inputs*

$$\begin{bmatrix} \dot{x}_1 \\ \dot{x}_2 \\ \dot{x}_3 \end{bmatrix} = \begin{bmatrix} -1.48 & 0.648 & -0.288 \\ -0.648 & -2.88 & 0.62 \\ 0.882 & -0.72 & -3.72 \end{bmatrix} \begin{bmatrix} g(x_1) \\ g(x_2) \\ g(x_3) \end{bmatrix}$$

$$+\begin{bmatrix} 0.4+\dfrac{1}{1+t} \\ 0.2+\dfrac{1}{1+t} \\ 0.3+\dfrac{1}{1+t} \end{bmatrix}, \tag{7.9}$$

*for $t \geq 0$, where*

$$g(s)=\frac{e^{s}-e^{-s}}{e^{s}+e^{-s}},\ s\in R.$$

Obviously, network (7.9) satisfies the conditions of Theorem 7.1, then it is ISC. Figure 7.1 shows the convergence of 50 trajectories of this network.

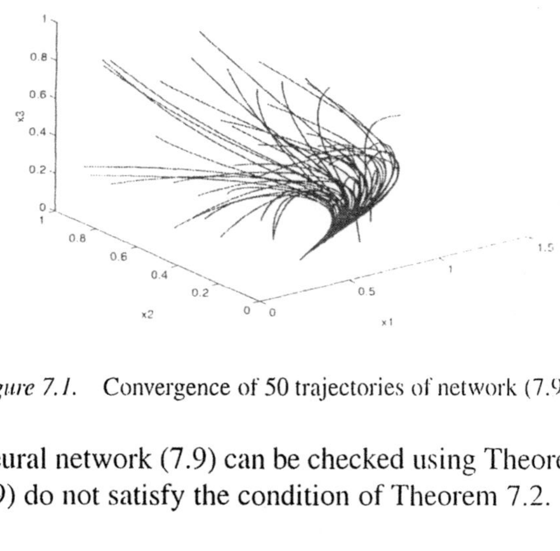


*Figure 7.1.* Convergence of 50 trajectories of network (7.9)

The ISC of neural network (7.9) can be checked using Theorem 7.2. In fact, the inputs of (7.9) do not satisfy the condition of Theorem 7.2.

EXAMPLE 7.2 *Consider the following neural network*

$$\begin{bmatrix} \dot{x}_1 \\ \dot{x}_2 \\ \dot{x}_3 \end{bmatrix} = \begin{bmatrix} -5 & 0 & 8 \\ -4 & -1 & 0 \\ 0 & 4 & -5 \end{bmatrix}\begin{bmatrix} f(x_1) \\ f(x_2) \\ f(x_3) \end{bmatrix} + \begin{bmatrix} 2+e^{-t} \\ -1+e^{-t} \\ 2+e^{-t} \end{bmatrix} \tag{7.10}$$

*for $t \geq 0$, where*

$$f(s)=\frac{1-e^{-s}}{1+e^{-s}},\ s\in R.$$

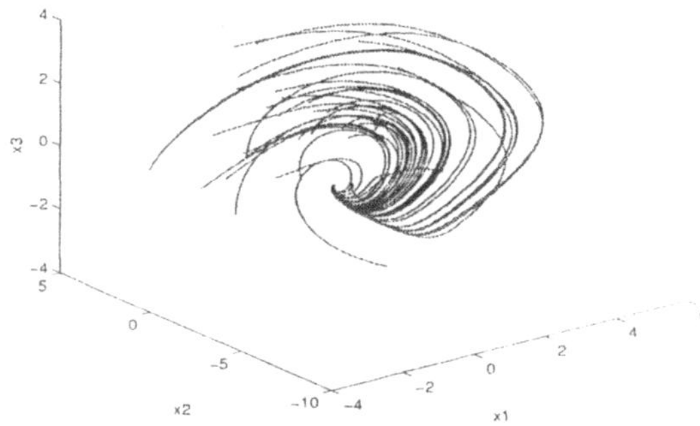


*Figure 7.2.* Convergence of 40 trajectories of network (7.10)

It can be checked that the network (7.10) satisfies the conditions of Theorem 7.2, thus it is ISC. Simulations for convergence of this network is given in Figure 7.2.

It is not difficult to see that the connection matrix of network (7.10) does not satisfy the condition of Theorem 7.1. Thus, its ISC can not be checked by Theorem 7.1. The above two examples show that Theorem 7.1 and Theorem 7.2 could be used in different situations.

# 2. RNN Model for Extracting Eigenvectors

## 2.1 Introduction

Computing eigenvectors of a matrix is an important and interesting problem in engineering, especially for computing eigenvectors corresponding to largest or smallest eigenvalues. There are many important applications of such computations, for example, applications in adaptive signal processing. Recently, there are many research works on this problem by using neural networks, say for examples [50, 87, 130, 131, 132, 138, 139, 140, 147, 148, 157, 183, 210, 214]. However, most of these works are centered on computing the eigenvectors of positive definite symmetric matrix corresponding to the largest or smallest eigenvalues. In this part, a more general case will be considered. A method based on the approach of neural networks to compute eigenvectors of any real symmetric matrix will be developed. This class of neural networks is described by differential equations, which is adapted from [130, 131, 132] and [157]. It is a model of recurrent neural networks that have asynchonous parallel processing ability and can achieve high computing performance.

Let $A$ be a $n \times n$ real symmetric matrix. The dynamics of neural network to be discussed is described as

$$\dot{X}(t) = X(t)^T X(t) A X(t) - X(t)^T A X(t) X(t), \qquad (7.11)$$

where $X \in R^n$ is the state of the network. It is a class of recurrent neural network model. Since equation (7.11) is a nonlinear differential equation, the dynamic behavior may not be evident. From the engineering point of view, network models that possess well understood dynamic behaviors are most attractive. This part clearly explores the dynamic behaviors of this network.

## 2.2 Equilibrium Points Analysis

$\xi \in R^n$ is an equilibrium point of (7.11) if and only if

$$\xi^T \xi A \xi - \xi^T A \xi \xi = 0.$$

Let $E$ denote the set of all equilibrium points of (7.11). Let $\lambda$ be an eigenvalue of $A$ and $V_\lambda$ be the eigenspace corresponding to $\lambda$.

THEOREM 7.3 *The set of all the equilibrium points of (7.11) is equal to the union of all eigenspace of A, that is*

$$E = \bigcup_\lambda V_\lambda.$$

*Proof*: For any $\xi \in \bigcup_\lambda V_\lambda$, if $\xi = 0$, then obviously $\xi \in E$. Suppose that $\xi \neq 0$, then there is an eigenvalue of $A$ such that $\xi \in V_\lambda$. That is $A\xi = \lambda\xi$, then

$$\xi^T \xi A \xi - \xi^T A \xi \xi = \xi^T \xi \lambda \xi - \lambda(\xi^T \xi)\xi = 0,$$

and so $\xi \in E$. This implies $E \supseteq \bigcup_\lambda V_\lambda$.

On the other hand, for any $\eta \in E$, we have $\eta^T \eta A \eta - \eta^T A \eta \eta = 0$. If $\eta = 0$, then $\eta \in \bigcup_\lambda V_\lambda$. If $\eta \neq 0$,

$$A\eta = \frac{\eta^T A \eta}{\eta^T \eta}\eta.$$

This shows that $\eta$ is an eigenvector of $A$. Then, $E \subseteq \bigcup_\lambda V_\lambda$. Therefore, we have $E = \bigcup_\lambda V_\lambda$. This completes the proof.

The above theorem shows that any equilibrium state of the network of (7.11) is an eigenvector of $A$ if it is not zero. Then, if the convergence of the network can be proven, it provides a method to compute the eigenvectors of $A$.

After an eigenvector is computed, it is easy to compute its corresponding eigenvalue. Suppose $v$ is an eigenvector of $A$, then the corresponding eigenvalue is $\lambda = \frac{v^T A v}{v^T v}$. In fact, we have

$$Av = \frac{v^T A v}{v^T v} v.$$

This provides a method to compute eigenvalues by known eigenvectors.

## 2.3 Representation of Solutions

In this subsection, the representation of the solutions of the network model of (7.11) will be discussed. It is clear that this network model is a nonlinear differential equation. Generally speaking, it is not easy to solve nonlinear differential equations. However, due to the fact that the matrix included in (7.11) is a symmetric matrix, this enables the representation of the solutions of (7.11) to be computed more easily.

Since $A$ is a symmetric matrix, then there exists an orthonormal basis of $R^n$ composed by eigenvectors of $R^n$. Let $\lambda_i (i = 1, \cdots, n)$ be eigenvalues of $A$ and $S_i (i = 1, \cdots, n)$ be the corresponding eigenvectors that compose an orthonormal basis of $A$. Then, for any $X \in R^n$, it can be represented as $X = \sum_{i=1}^{n} z_i S_i$, where $z_i (i = 1, \cdots, n)$ are some constants.

THEOREM 7.4 *For any $X(0) \in R^n$, if $X(0) = \sum_{i=1}^{n} z_i(0) S_i \neq 0$. Then the solution of (1) starting from $X(0)$ can be represented as*

$$X(t) = \sum_{i=1}^{n} \sqrt{\frac{X(0)^T X(0)}{\sum_{j=1}^{n} z_j^2(0) e^{2X(0)^T X(0)(\lambda_j - \lambda_i)t}}} z_i(0) S_i$$

*for all $t \geq 0$.*

*Proof*: For any $t \geq 0$, let $X(t)$ be the solution of (7.11) starting from $X(0)$. Then, it follows

$$\begin{aligned} \frac{d[X(t)^T X(t)]}{dt} &= 2X(t)\dot{X}(t) \\ &= X(t)^T X(t) X(t)^T A X(t) - X(t)^T A X(t) X(t)^T X(t) \\ &= 0 \end{aligned}$$

and so $X(t)^T X(t) = X(0)^T X(0)$ for all $t \geq 0$.

Rewrite (7.11) as

$$\dot{X}(t) = X(0)^T X(0) A X(t) - X(t)^T A X(t) X(t). \tag{7.12}$$

Let $X(t)$ be represented under the basis of $S_i (i = 1, \cdots, n)$ as

$$X(t) = \sum_{i=1}^{n} z_i(t) S_i.$$

Then, from (7.12), it follows

$$\dot{z}_i(t) = X(0)^T X(0) \lambda_i z_i(t) - \left( \sum_{j=1}^{n} \lambda_j z_j^2(t) \right) z_i(t) \tag{7.13}$$

for $t \geq 0$ and all $i = 1, \cdots, n$.

Since $X(0) \neq 0$, then there exists a $r(1 \leq r \leq n)$ such that $z_r(0) \neq 0$, it follows from (7.13) that

$$z_r(t) = z_r(0)e^{\int_0^t \left[X(0)^T X(0)\lambda_r - \left(\sum_{j=1}^n \lambda_j z_j^2(\theta)\right)\right] d\theta} \neq 0$$

for all $t \geq 0$. Without lose of generality, $z_r(0) > 0$, then it follows $z_r(t) > 0$ for all $t \geq 0$.

It is easy to obtain from (7.13) that

$$\frac{d}{dt}\left[\frac{z_j(t)}{z_r(t)}\right] = X(0)^T X(0)(\lambda_j - \lambda_r)\left[\frac{z_j(t)}{z_r(t)}\right], (j = 1, \cdots, n),$$

for all $t \geq 0$. Then,

$$\frac{z_j(t)}{z_r(t)} = \frac{z_j(t)}{z_r(t)} e^{X(0)^T X(0)(\lambda_j - \lambda_r)}, (j = 1, \cdots, n), \tag{7.14}$$

for all $t \geq 0$. Next, $z_r(t)$ will be computed.

From (7.13),

$$\frac{dz_r(t)}{dt} = X(0)^T X(0)\lambda_r z_r(t) - \left(\sum_{j=1}^n \lambda_j z_j^2(t)\right) z_r(t)$$

then

$$\frac{d}{dt}\left[\frac{1}{z_r^2(t)}\right] = -2\lambda_r X(0)^T X(0)\left[\frac{1}{z_r^2(t)}\right] + 2\sum_{j=1}^n \lambda_j \left[\frac{z_j(t)}{z_r(t)}\right]^2.$$

Substituting (7.14) into the right hand side of the above equation, we get

$$\frac{d}{dt}\left[\frac{1}{z_r^2(t)}\right] = -2\lambda_r X(0)^T X(0)\left[\frac{1}{z_r^2(t)}\right] + 2\sum_{j=1}^n \lambda_j \left[\frac{z_j(0)}{z_r(0)}\right]^2 e^{2(\lambda_j - \lambda_r)t}$$

Integrating the equation, it follows that

$$\begin{aligned}
\frac{1}{z_r^2(t)} &= \frac{1}{z_r^2(0)} e^{-2X(0)^T X(0)\lambda_r t} \\
&+ 2\sum_{j=1}^n \lambda_j \left[\frac{z_j(0)}{z_r(0)}\right]^2 \int_0^t e^{-2X(0)^T X(0))(\lambda_r t - \lambda_j s)} ds \\
&= \frac{1}{z_r^2(0) X(0)^T X(0)} \sum_{j=1}^n z_j^2(0) e^{2X(0)^T X(0)(\lambda_j - \lambda_r)t.}
\end{aligned}$$

Thus

$$z_r(t) = \sqrt{\frac{X(0)^T X(0)}{\sum_{j=1}^{n} z_j^2(0) e^{2X(0)^T X(0)(\lambda_j - \lambda_r)t}}} z_r(0). \tag{7.15}$$

Substituting (7.15) into (7.14),

$$\begin{aligned} z_i(t) &= z_r(t) \frac{z_i(0)}{z_r(0)} e^{X(0)^T X(0)(\lambda_i - \lambda_r)t} \\ &= \sqrt{\frac{X(0)^T X(0)}{\sum_{j=1}^{n} z_j^2(0) e^{2X(0)^T X(0)(\lambda_j - \lambda_i)t}}} z_i(0) \end{aligned}$$

for all $t \geq 0$ and $i = 1, \cdots, n$. Therefore,

$$X(t) = \sum_{i=1}^{n} \sqrt{\frac{X(0)^T X(0)}{\sum_{j=1}^{n} z_j^2(0) e^{2X(0)^T X(0)(\lambda_j - \lambda_i)t}}} z_i(0) S_i$$

for all $t \geq 0$. This completes the proof.

Theorem 7.4 above shows the solutions of the networks can be represented in terms of a set of orthogonal eigenvectors. This is quite convenient for us to study the convergence behaviors of the network.

## 2.4 Convergence Analysis

In this subsection, the convergence of the solutions of the neural network of (7.11) will be considered. For an artificial neural network of continuous model, the convergence property is crucial. It is a basic requirement for designing a neural network to guarantee convergence. In the last section, the representation of the solutions of network (7.11) has been established. Using this representation, the convergence property can be studied on a sound foundation.

THEOREM 7.5 *Each solution of (7.11) starting from any nonzero points in $R^n$ will converge to an eigenvector of $A$.*

*Proof*: Let $\lambda_i (i = 1, \cdots, n)$ be all the eigenvalues of $A$ ordered by $\lambda_1 \geq \cdots \geq \lambda_n$. Suppose that $S_i (i = 1, \cdots, n)$ is an orthonormal basis in $R^n$ such that each $S_i$ is an eigenvector of $A$ corresponding to the eigenvalue $\lambda_i$. Let $\sigma_i (i = 1, \cdots, m)$ be all the distinct eigenvalues of $A$ ordered by $\sigma_1 > \cdots > \sigma_m$. For any $i$, $1 \leq i \leq m$, denote the algebraic sum of the multiplicity of $\sigma_1, \cdots, \sigma_i$

by $k_i$. Obviously, we have $k_m = n$. For convenience, we denote $k_0 = 1$. It is easy to see that $\lambda_i = \sigma_r$ for all $i \in [k_{r-1}, k_r]$, and $S_i \in V_{\sigma_r}$ for all $i \in [k_{r-1}, k_r]$.

Let $X(0)$ be a nonzero point in $R^n$ and suppose that $X(0) = \sum_{i=1}^n z_i(0)S_i$. Define

$$l = \min\{l | 1 \leq l \leq n, z_l(0) \neq 0\}$$

then there exists a $r \in \{1, \cdots, m\}$ such that $k_{r-1} \leq l \leq k_r$. By Theorem 7.4, the solution of (7.11) starting from $X(0)$ must satisfy

$$\begin{aligned} X(t) &= \sum_{i=1}^n \sqrt{\frac{X(0)^T X(0)}{\sum_{j=1}^n z_j^2(0) e^{2X(0)^T X(0)(\lambda_j - \lambda_i)t}}} z_i(0) S_i \\ &= \sqrt{\frac{X(0)^T X(0)}{\sum_{j=l}^{k_r} z_j^2(0) + \sum_{j=k_r+1}^n z_j^2(0) e^{2X(0)^T X(0)(\lambda_j - \sigma_r)t}}} \sum_{i=l}^{k_r} z_i(0) S_i \\ &+ \sum_{i=k_r+1}^n \Pi_i(t) \end{aligned}$$

for $t \geq 0$, where each $\Pi_i(t)$ equal to

$$\sqrt{\frac{X(0)^T X(0)}{\sum_{j=l}^{k_r} z_j^2(0) e^{2X(0)^T X(0)(\sigma_r - \lambda_i)t} + \sum_{j=k_r+1}^n z_j^2(0) e^{2X(0)^T X(0)(\lambda_j - \lambda_i)t}}} z_i(0) S_i.$$

Clearly, $\Pi_i(t) \to 0$ as $t \to +\infty$. Then, we have

$$X(t) \to \sqrt{\frac{X(0)^T X(0)}{\sum_{j=l}^{k_r} z_j^2(0)}} \sum_{i=l}^{k_r} z_i(0) S_i \in V_{\sigma_r}$$

as $t \to +\infty$. This completes the proof.

THEOREM 7.6 *For any nonzero point $X(0) \in R^n$, if $X(0)$ is not orthogonal to $V_{\sigma_1}$, then the solution of (7.11) starting from $X(0)$ converges to an eigenvector corresponding to the largest eigenvalue.*

*Proof:* Suppose $X(0) = \sum_i^n z_i(0)S_i$, then $X(0)$ is not orthogonal to $V_{\sigma_1}$ if and only if there exists a $l(1 \leq l \leq k_1)$ such that $z_l(0) \neq 0$. From the proof of theorem 7.5, it follows that

$$X(t) \rightarrow \sqrt{\frac{X(0)^T X(0)}{\sum_{j=l}^{k_1} z_j^2(0)}} \sum_{i=l}^{k_1} z_i(0) S_i \in V_{\sigma_1},$$

as $t \rightarrow +\infty$. This completes the proof.

The above theorem gives the necessary and sufficient conditions for the network to converge to eigenvectors corresponding to the largest eigenvalue. These conditions depend on $V_{\sigma_1}$. As $V_{\sigma_1}$ is not known a prior, it is not practical to choose initial values that are always orthogonal to $V_{\sigma_1}$ in advance. However, from the fact that the dimension of $V_{\sigma_1}$ is always less than the dimension of $R^n$, the problem can be solved by giving the initial values some random perturbations. In this way, a high probability of having initial values that are not orthogonal to $V_{\sigma_1}$ can be ensured. Simulations in next section will further show this works well.

Theorem 7.6 shows how to compute eigenvector corresponding to the largest eigenvalue. It is also very interesting to compute eigenvector corresponding to the smallest eigenvalue. The next theorem provides such a method.

THEOREM 7.7 *Replacing $A$ in (7.11) with $-A$, and suppose $X(0)$ is a nonzero vector in $R^n$ which is not orthogonal to $V_{\sigma_m}$, then the solution starting from $X(0)$ will converge to an eigenvector corresponding to the smallest eigenvalue of $A$.*

*Proof*: Noting the fact that $-\sigma_m$ is the largest eigenvalue of $-A$, by applying Theorem 7.6, it follows that the solution of (7.11) starting from $X(0)$ will converge to an eigenvector $v$ corresponding to the largest eigenvalue $-\sigma_m$ of $-A$. Then, $-Av = -\sigma_m v$, that is $Av = \sigma_m v$. This shows that $v$ is actually an eigenvector corresponding to the smallest eigenvalue $\sigma_m$ of $A$. The proof is completed.

Theorem 7.7 shows that to compute eigenvector corresponding to the smallest eigenvalue of $A$, it can just simply replace $A$ in the network (7.11) with $-A$.

THEOREM 7.8 *If $X(0)$ is orthogonal to each $S_i (i = 1, \cdots, c)$, where $c \leq n$ is a constant, then the solution of (7.11) starting from $X(0)$ converges to an eigenvector which is also orthogonal to each $S_i (i = 1, \cdots, n)$.*

*Proof*: Let $X(0)$ be represented under the basis $S_i (i = 1, \cdots, n)$ as $X(0) = \sum_{i=1}^{n} z_i(0) S_i$. Since $X(0)$ is orthogonal to each $S_i (i = 1, \cdots, c)$, then $z_i(0) = 0 (i = 1, \cdots, c)$. Obviously, we have

$$c \leq l = \min\{i | 1 \leq i \leq n, z_i(0) \neq 0\}.$$

By the proof of Theorem 7.5, the solution starting from $X(0)$ converges to

$$\sqrt{\frac{X(0)^T X(0)}{\sum_{j=l}^{k_r} z_j^2(0)}} \sum_{i=l}^{k_r} z_i(0) S_i \in V_{\sigma_r}$$

which is orthogonal to each $S_i (i = 1, \cdots, n)$. This completes the proof.

## 2.5 Examples and Simulation Results

In this subsection, some simulation results will be given to illustrate the ability of the proposed network.

Let $Q$ be any real matrix, it can be randomly generated, then we define

$$A = \frac{Q^T + Q}{2}$$

which is clearly a symmetric matrix.

In the first example, the $5 \times 5$ input matrix $A$ is:

$$\begin{bmatrix} 0.7663 & 0.4283 & -0.3237 & -0.4298 & -0.1438 \\ 0.4283 & 0.2862 & 0.0118 & -0.2802 & 0.1230 \\ -0.3237 & 0.0118 & -0.9093 & -0.4384 & 0.7684 \\ -0.4298 & -0.2802 & -0.4384 & -0.0386 & -0.1315 \\ -0.1438 & 0.1230 & 0.7684 & -0.1315 & -0.4480 \end{bmatrix}.$$

Using the network model, the estimated eigenvectors corresponding to $\lambda_{max}$ and $\lambda_{min}$ respectively are:

$$\xi_{max} = \begin{bmatrix} 1.0872 \\ 0.6264 \\ -0.0809 \\ -0.4736 \\ -0.0472 \end{bmatrix}, \xi_{min} = \begin{bmatrix} 0.1882 \\ 0.0600 \\ 1.3209 \\ 0.3697 \\ -0.8446 \end{bmatrix}.$$

Applying the model on $A$ leads to Figure 7.3(a). The result is an estimation of the eigenvector, that is $\xi_{max}$ above, corresponding to the largest eigenvalue. An immediate result follows is an estimation to the maximum eigenvalue of 1.2307, which is an accurate approximation to the true value with a precision of 0.0001. For reader's reference, the true maximum and minimum eigenvalues as computed by Matlab are shown here:

| | | | |
|---|---|---|---|
| maximum eigenvalue | $\lambda_{max}$ | = | 1.2307 |
| minimum eigenvalue | $\lambda_{min}$ | = | -1.5688 |

Another advantage of the technique is that it allows the calculation of an eigenvector corresponding to the minimum eigenvalue. This is due to Theorem

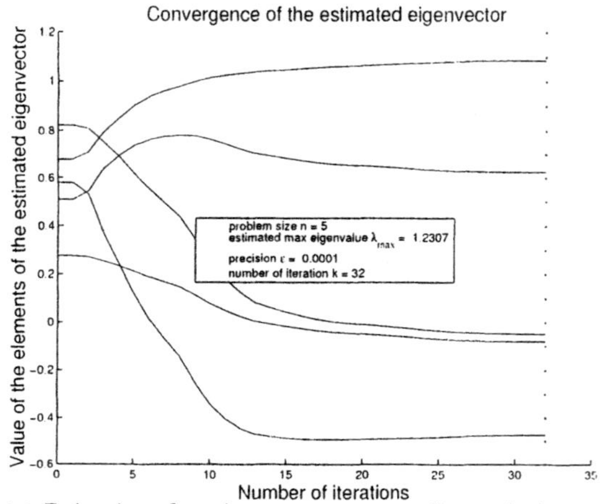


(a) Estimation of an eigenvector corresponding to the largest eigenvalue.

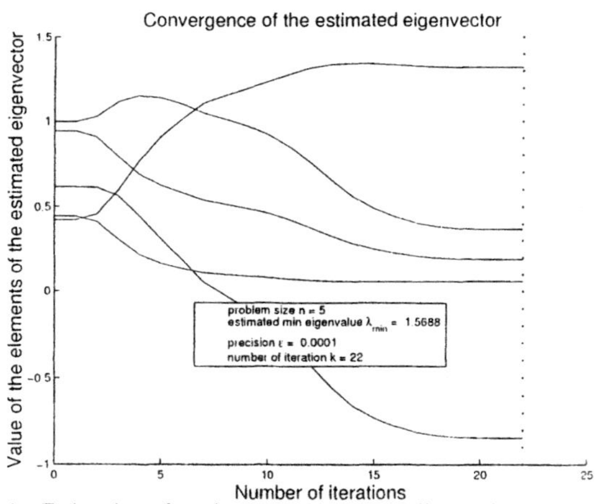


(b) Estimation of an eigenvector corresponding to the smallest eigenvalue.

*Figure 7.3.* Results of estimating the largest and smallest eigenvalue and the corresponding eigenvectors of the matrix $A$.

7.7. The result is shown in Figure 7.3(b). By feeding the model with $-A$, it get that $\xi_{min}$, an estimation to the desired eigenvector, as well as the magnitude of the smallest eigenvalue, 1.5688, which is an accurate estimation just with the sign flipped. The generated eigenvector also corresponds to the target eigenvalue.

Consider another example with a $7 \times 7$ input matrix $B$:

$$\begin{bmatrix} 0.6037 & 0.5587 & -0.3907 & 0.1629 & -0.2079 & 0.3688 & 0.3990 \\ 0.5587 & 0.9918 & -0.1018 & 0.4110 & 0.1256 & -0.2908 & -0.0014 \\ -0.3907 & -0.1018 & -0.3977 & -0.2967 & 0.0273 & -0.3578 & 0.2106 \\ 0.1629 & 0.4110 & -0.2967 & -0.9101 & 0.6153 & 0.3519 & -0.3277 \\ -0.2079 & 0.1256 & 0.0273 & 0.6153 & -0.8451 & -0.0503 & -0.7252 \\ 0.3688 & -0.2908 & -0.3578 & 0.3519 & -0.0503 & 0.3475 & 0.1201 \\ 0.3990 & -0.0014 & 0.2106 & -0.3277 & -0.7252 & 0.1201 & -0.9565 \end{bmatrix}$$

Using the network model, the estimated eigenvectors corresponding to $\lambda_{max}$ and $\lambda_{min}$ respectively are:

$$\xi_{max} = \begin{bmatrix} 1.1439 \\ 1.3386 \\ -0.3907 \\ 0.3747 \\ 0.0256 \\ 0.2602 \\ 0.1042 \end{bmatrix}, \xi_{min} = \begin{bmatrix} 0.0602 \\ -0.0392 \\ 0.2305 \\ 0.4682 \\ -0.8533 \\ -0.0371 \\ -0.6757 \end{bmatrix}.$$

Similarly, accurate estimations are produced using the model, for both eigenvectors corresponding to the maximum and minimum eigenvalues. Figure 7.4 shows the convergence of the eigenvectors estimation of the network for both cases. The true maximum and minimum eigenvalues as computed by Matlab are also listed below:

| | | | |
|---|---|---|---|
| maximum eigenvalue | $\lambda_{max}$ | = | 1.5599 |
| minimum eigenvalue | $\lambda_{min}$ | = | -1.7461 |

For both simulations, the desired precision is set to 0.0001. This accuracy level can be changed according to the need of the user. The network converge up to this predefined accuracy successfully, as shown theoretically and practically. These simulations further confirm that the network model can be used to compute the eigenvectors and eigenvalues of symmetrical matrix.

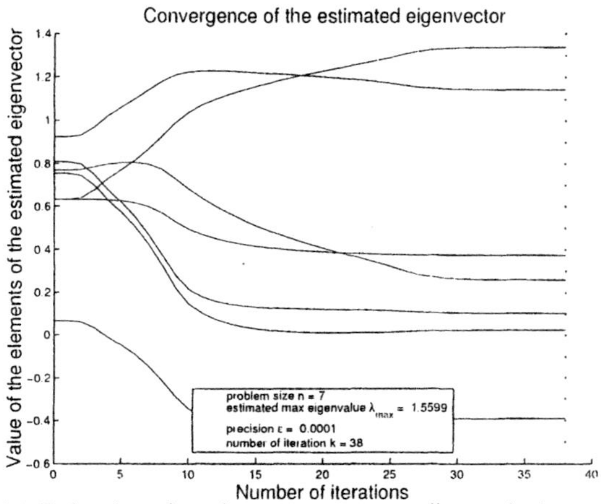


(a) Estimation of an eigenvector corresponding to the largest eigenvalue.

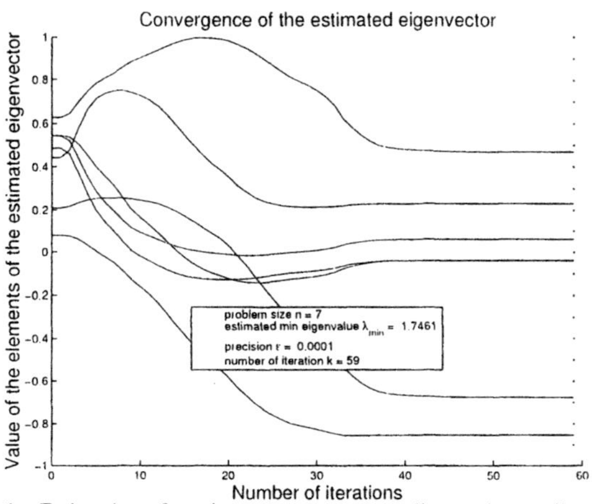


(b) Estimation of an eigenvector corresponding to the smallest eigenvalue.

*Figure 7.4.* Results of estimating the largest and smallest eigenvalue and the corresponding eigenvectors of the matrix $B$.

## Chapter 8

# DISCRETE RECURRENT NEURAL NETWORKS

## 1. Introduction

Generally, a discrete time RNN can be described by the difference system model

$$x(k+1) = F(x(k)), \quad k = 0, 1, 2, \cdots \tag{8.1}$$

where $x \in R^n$, $F : R^n \to R^n$ is a mapping which can be bounded or unbounded. By suitably selecting the mapping $F$ in (8.1), various classes of discrete RNNs can be obtained.

Discrete time RNNs possess advantages for direct computer simulations over digital simulations of continuous time RNN, and they can be easily implemented using digital hardware. Discrete time neural networks have been widely studied in recent years by many authors. In [84, 71, 27], Hopfield discrete neural networks and discrete CNNs are applied to associative memory applications. In [38, 166, 150], discrete neural networks are used for image processing. In [126], discrete RNNs are used for digitial filters. In [112, 153, 206] discrete RNNs are used to some optimization problems. In [217, 214], discrete RNNs are used to address winner-take-all problem. In [89] and [154], discrete neural networks of the brain-state-in-box model were studied. General dynamic properties of DNNs can be found in [27, 55, 88, 90, 91, 126, 142, 166].

In Section 2, a model of discrete RNNs with unsaturating piecewise linear activation functions will be studied. In particular, to the multistability of this discrete RNN model will be considered. It is known that there are many applications where multi-stable dynamics is essential for neural networks [51, 75, 76, 203] to be used effectively. Multi-stability analysis for discrete RNNs is different from mono-stability analysis in general. In mono-stability analysis, the objective is to derive conditions that guarantee each network contains only one equilibrium point, and all the trajectories of the network converge to it. In multi-stability analysis, the networks are allowed to have multiple equilibrium points. Stable and unstable equilibrium points, and even periodic trajectories may co-exist in a network. Generally, the usual global stability conditions are not adequately applicable to multi-stable networks. Three problems will be addressed: boundedness, global attractivity and complete convergent.

In Section 3, a model of discrete time RNNs applied to the problem of winner-take-all will be presented. The dynamic properties of the network will be discussed in detail and extensive simulations will be given to demonstrate the ability of it for winner-take-all problem.

## 2. Discrete RNNs with Unsaturating Piecewise Linear Activation Functions

Consider the following model of discrete RNNs with unbounded activation functions

$$x_i(k+1) = \sum_{j=1}^{n} w_{ij}\sigma(x_j(k)) + h_i, \quad (i = 1, \cdots, n), \tag{8.2}$$

or its equivalent vector form,

$$x(k+1) = W\sigma(x(k)) + h,$$

for all $k \geq 0$, where each $x_i$ denotes the activity of neuron $i$, $W = (w_{ij})_{n \times n}$ is a real $n \times n$ matrix, each of its elements $w_{ij}$ denotes the synaptic weights and represents the strength of the synaptic connection from neuron $j$ to neuron $i$, $x = (x_1, \cdots, x_n)^T \in R^n$, $h \in R^n$ denotes external inputs. The activation function $\sigma$ is defined as follows:

$$\sigma(s) = \max\{0, s\}, \quad s \in R.$$

For any vector $x \in R^n$, denote

$$\sigma(x) = (\sigma(x_1), \cdots, \sigma(x_n))^T \in R^n.$$

The activation function $\sigma(\cdot)$ is an unsaturating piecewise linear function, which is continuous, unbounded.

The boundedness, global attractivity and complete convergence of the network (8.2) will be addressed. They are important issues to be addressed in the multi-stability analysis of neural networks. Conditions will be derived to guarantee boundedness, and also allow the network to have multi-equilibrium points. This will be achieved by using local inhibition to the network. Under such non-divergence conditions, it will show that the network will possess compact sets which attract all the trajectories of the network. Explicit inequalities for locating the attracting compact sets will be given. In addition, if the weight matrix of the network possesses a certain symmetry, complete stability of the network (8.2) will be proved.

## 2.1 Preliminaries

In this section, some preliminaries which will be used in the following to establish our theory will be provided.

LEMMA 8.1 *Given any constant* $c \geq 0$,

$$c \cdot \sigma(s) = \sigma(c \cdot s)$$

*for all* $s \in R$.

*Proof*: Since $c \geq 0$,

$$c \cdot \sigma(s) = c \cdot \max\{0, s\} = \max\{0, c \cdot s\} = \sigma(c \cdot s).$$

This completes the proof.

LEMMA 8.2 *Given any constants* $s$ *and* $\theta$, *if* $s \leq \theta$, *then*

$$\sigma(s) \leq \sigma(\theta).$$

*Proof*: Since $s \leq \theta$,

$$\max\{0, s\} \leq \max\{0, \theta\}.$$

Then, $\sigma(s) \leq \sigma(\theta)$. The proof is completed.

LEMMA 8.3 *It holds that*

$$\int_v^u [\sigma(s) - \sigma(v)]\, ds \geq \frac{[\sigma(u) - \sigma(v)]^2}{2}$$

*for all* $u, v \in R$.

*Proof*: Case 1: $u, v \geq 0$. Then, we have

$$\begin{aligned} \int_v^u [\sigma(s) - \sigma(v)]\, ds &= \int_v^u [s - v]\, ds \\ &= \frac{[\sigma(u) - \sigma(v)]^2}{2}. \end{aligned}$$

Case 2: $u, v \leq 0$. Then,

$$\int_v^u [\sigma(s) - \sigma(v)]\, ds = 0 = \frac{[\sigma(u) - \sigma(v)]^2}{2}.$$

Case 3: $u \geq 0$, $v \leq 0$. Then,

$$\begin{aligned} \int_v^u [\sigma(s) - \sigma(v)]\, ds &= \int_0^u s ds \\ &= \frac{[\sigma(u) - \sigma(v)]^2}{2}. \end{aligned}$$

Case 4: $u \leq 0$, $v \geq 0$. Then,

$$\begin{aligned}\int_v^u [\sigma(s) - \sigma(v)]\, ds &= \int_v^0 [s - \sigma(v)]\, ds - \int_0^u \sigma(v) ds \\ &= \frac{[\sigma(u) - \sigma(v)]^2}{2} - u\sigma(v) \\ &\geq \frac{[\sigma(u) - \sigma(v)]^2}{2}.\end{aligned}$$

The proof is completed.

Denote in this paper a norm in $R^n$ by

$$\|x\| = \sqrt{\sum_{i=1}^{n} |x_i|^2},$$

for any $x \in R^n$.

DEFINITION 8.1 *The network (8.2) is said to be bounded if each of its trajectories is bounded.*

DEFINITION 8.2 *Let $S$ be a compact subset of $R^n$. We denote the $\epsilon$ − neighborhood of $S$ by $S_\epsilon$. The compact set $S$ is said to globally attract the network (8.2), if for any $\epsilon > 0$, all trajectories of (8.2) ultimately enter and remain in $S_\epsilon$.*

Obviously, if every trajectory $x(t)$ of (8.2) satisfy

$$\begin{cases} \lim_{t \to +\infty} \sup x(t) \in S \\ \lim_{t \to +\infty} \inf x(t) \in S \end{cases}$$

then, $S$ globally attracts the network (8.2).

A vector $x^* \in R^n$ is called an equilibrium point of (8.2), if it satisfies

$$-x^* + W\sigma(x^*) + h \equiv 0.$$

Denote by $\Omega$ to the set of equilibrium points of (8.2).

DEFINITION 8.3 *The network (8.2) is said to be completely convergent, if each trajectory $x(k)$ of (8.2) satisfies*

$$dist\,(x(k), \Omega) \triangleq \min_{x^* \in \Omega} \| x(k) - x^* \| \longrightarrow 0$$

*as $k \to +\infty$.*

Throughout this chapter, for any constant $c \in R$, denote $c^+ = \max(0, c)$, $c^- = \min(0, c)$. Obviously, $c^+ \geq 0$, $c^- \leq 0$. For any matrix $A = (a_{ij})_{n \times m}$, we denote $A^+ = \left(a_{ij}^+\right)$, where

$$a_{ij}^+ = \max(0, a_{ij}), (i, j = 1, \cdots, n).$$

## 2.2 Boundedness and Global Attractivity

Since the activation function $\sigma(\cdot)$ is unbounded, the trajectories of (8.2) may diverge. For example, consider the simple one-dimensional network

$$x(k+1) = \sigma(x(k)) + 1,$$

for $k \geq 0$. Let $x(0) = 0$, then the solution starting from this point is $x(k) = k$ for $k \geq 0$. Clearly, it is unbounded. Networks with a divergent property are not amenable to applications. Boundedness is a basic property for any applicable neural networks.

Using the method of local inhibition [203], in this section, conditions will be derived to guarantee boundedness. Due to the advantage of using local inhibition, these conditions allow the network (8.2) to have multi-equilibrium points. Moreover, explicit expressions will be provided, locating the compact sets which globally attract all the trajectories of the network (8.2).

THEOREM 8.1 *If there exist constants $\alpha_i > 0 (i = 1, \cdots, n)$ such that*

$$\frac{1}{\alpha_i}\sum_{j=1}^{n}\alpha_j w_{ij}^{+} < 1, \quad (i = 1, \cdots, n),$$

*then the network (8.2) is bounded, and the compact set*

$$S = \{x \,|a_i \leq x_i \leq b_i, 1 \leq i \leq n\}$$

*globally attracts the network (8.2), where*

$$\begin{cases} b_i = \dfrac{\alpha_i}{1-\gamma} \cdot \max\limits_{1\leq j\leq n}\left\{\dfrac{h_j^{+}}{\alpha_j}\right\} \\ a_i = \displaystyle\sum_{j=1}^{n} w_{ij}^{-} \cdot b_j + h_i^{-} \\ \gamma = \max\limits_{1\leq i\leq n}\left\{\dfrac{1}{\alpha_i}\displaystyle\sum_{j=1}^{n}\alpha_j w_{ij}^{+}\right\}. \end{cases}$$

*Proof*: Since $\sigma(\cdot) \geq 0$, it follows from (8.2) that

$$x_i(k+1) \leq \sum_{j=1}^{n} w_{ij}^{+}\sigma(x_j(k)) + h_i^{+}, (i = 1, \cdots, n) \quad (8.3)$$

for $k \geq 0$.

Define

$$v_i(k) = \frac{x_i(k)}{\alpha_i}, (i = 1, \cdots, n),$$

for $k \geq 0$. Then, from (8.3) and Lemma 8.1,

$$v_i(k+1) \leq \frac{1}{\alpha_i}\left(\sum_{j=1}^{n} \alpha_j w_{ij}^{+}\sigma(v_j(k)) + h_i^{+}\right), (i = 1, \cdots, n) \tag{8.4}$$

for $k \geq 0$.

By Lemma 8.2 and (8.4),

$$\sigma(v_i(k+1)) \leq \frac{1}{\alpha_i}\left(\sum_{j=1}^{n} \alpha_j w_{ij}^{+}\sigma(v_j(k)) + h_i^{+}\right), (i = 1, \cdots, n) \tag{8.5}$$

for $k \geq 0$.

Define

$$V(k) = \max_{1 \leq j \leq n} \{\sigma(v_j(k))\} \tag{8.6}$$

for $k \geq 0$.

From (8.4)

$$\begin{aligned} \sigma(v_i(k+1)) &\leq \left(\frac{1}{\alpha_i}\sum_{j=1}^{n} \alpha_j w_{ij}^{+}\right) \cdot V(k) + \frac{h_i^{+}}{\alpha_i} \\ &\leq \gamma \cdot V(k) + \max_{1 \leq j \leq n}\left\{\frac{h_j^{+}}{\alpha_j}\right\} \end{aligned}$$

and so

$$V(k+1) \leq \gamma \cdot V(k) + \max_{1 \leq j \leq n}\left\{\frac{h_j^{+}}{\alpha_j}\right\}$$

for $k \geq 0$. Then

$$\begin{aligned} V(k+1) &\leq \gamma^{k+1} \cdot V(0) + \max_{1 \leq j \leq n}\left\{\frac{h_j^{+}}{\alpha_j}\right\} \cdot \sum_{l=0}^{k} \gamma^l \\ &\leq \gamma^{k+1} \cdot V(0) + \frac{1}{1-\gamma} \cdot \max_{1 \leq j \leq n}\left\{\frac{h_j^{+}}{\alpha_j}\right\} \end{aligned}$$

for $k \geq 0$. Thus, by (8.6)

$$\sigma(x_i(k+1)) \leq \alpha_i \gamma^{k+1} \cdot V(0) + \frac{\alpha_i}{1-\gamma} \cdot \max_{1 \leq j \leq n}\left\{\frac{h_j^{+}}{\alpha_j}\right\}$$

for $k \geq 0$. Clearly,

$$\lim_{k \to +\infty} \sup x_i(k+1) \leq \lim_{k \to +\infty} \sup \sigma(x_i(k+1))$$

$$\begin{aligned} &\leq \frac{\alpha_i}{1-\gamma} \cdot \max_{1 \leq j \leq n} \left\{ \frac{h_j^+}{\alpha_j} \right\} \\ &= b_i, (i = 1, \cdots, n). \end{aligned}$$

On the other hand, from (8.2),

$$x_i(k+1) \geq \sum_{j=1}^{n} w_{ij}^- \sigma(x_j(k)) + h_i^-, (i = 1, \cdots, n)$$

for $k \geq 0$. Noting that $w_{ij}^- \leq 0$, it follows that

$$\begin{aligned} \lim_{k \to +\infty} \inf x_i(k+1) &\geq \sum_{j=1}^{n} w_{ij}^- \cdot \lim_{k \to +\infty} \sup \sigma(x_j(k)) + h_i^- \\ &\geq \sum_{j=1}^{n} w_{ij}^- \cdot b_j + h_i^- \\ &= a_i. \end{aligned}$$

This shows that the compact set $S$ globally attracts any trajectory of (8.2). The proof is completed.

In Theorem 8.1, if $h_i \leq 0$ $(i = 1, \cdots, n)$, then $b_i = 0 (i = 1, \cdots, n)$, the compact set become

$$S = \left\{ x \middle| h_i^- \leq x_i \leq 0, (i = 1, \cdots, n) \right\}.$$

In particular, if $h_i = 0$ $(i = 1, \cdots, n)$, the attracting sets contain only the origin, i.e., the network (8.2) can have only the origin as its unique equilibrium point and all trajectories of (8.2) converge to the origin.

## 2.3 Complete Convergence

In this section, the complete convergence of the network (8.2) will be studied. Complete convergence means that every trajectory converges to the equilibrium set of the network. It allows a network to possess multi-equilibrium points.

Since the activation function $\sigma(\cdot)$ in the network (8.2) is an unsaturating piecewise function, the trajectories of (8.2) may diverge. As it is well known that the property of bounded trajectories is a necessary condition for convergence, the non-divergence conditions given in last section can be applied first.

Define a set

$$S = \left\{ \begin{bmatrix} e_1 & 0 & \cdots & 0 \\ 0 & e_2 & \cdots & 0 \\ \vdots & \vdots & \cdots & \vdots \\ 0 & 0 & \cdots & e_n \end{bmatrix} \middle| e_i = 0 \text{ or } 1, 1 \leq i \leq n \right\}.$$

Clearly, each elements of $S$ is a $n \times n$ matrix and the set $S$ has total $2^n$ elements. Using this notation, a representation of each trajectory of the network (1) can be given.

LEMMA 8.4 *Given any initial condition $x(0) \in R^n$, the trajectory of (1) starting from $x(0)$ can be represented as*

$$x(k+1) = \left[\prod_{j=1}^{k} WE(k-j)\right] x(0) + \sum_{i=0}^{k-1} \left[\prod_{j=0}^{i} WE(k-j)\right] h + h$$

*for all $k \geq 0$, where each $E(i) \in S$. Moreover, every component of $x(k+1)$ must be a polynomial with order no more than $k$.*

*Proof*: At each step $k$, there exists a matrix $E(k) \in S$ such that $\sigma(x(k)) = E(k)x(k)$, then (8.2) can be rewritten as

$$x(k+1) = WE(k)x(k) + h$$

for $k \geq 0$. By iterating, the result follows.

Since the set $S$ has $2^n$ elements, from the trajectory representation, each component of $x(k+1)$ must be a polynomial with order at most $k$. This completes the proof.

THEOREM 8.2 *Suppose the network (8.2) is bounded. If there exists a diagonal positive definite matrix $D$ such that $D(I+W)$ is a symmetric positive definite matrix, then the network (8.2) is completely convergent.*

*Proof*: For any $x_0 \in R^n$, let $x(k)$ be the solution of (8.2) starting from $x_0$. Suppose $D = diag(d_1, \cdots, d_n)$, clearly, $d_i > 0 (i = 1, \cdots, n)$.

Constructing the following energy function

$$\begin{aligned} E(x(k)) &= -\frac{1}{2}\sigma^T(x(k))DW\sigma(x(k)) \\ &\quad -\sigma^T(x(k))Dh + \sigma^T(x(k))Dx(k) \\ &\quad -\sum_{i=1}^{n} d_i \int_0^{x_i(k)} \sigma(s)ds \end{aligned} \tag{8.7}$$

for all $k \geq 0$.

For notation convenience, denote

$$z(k+1) = \sigma(x(k+1)) - \sigma(x(k))$$

for $k \geq 0$.

From (8.7) and (8.2),

$$\begin{aligned}
E(x(k+1)) - E(x(k)) &= -\frac{1}{2} z^T(k+1) D W z(k+1) \\
&\quad - z^T(k+1) D \left[ W \sigma(x(k)) + h \right] \\
&\quad + \sigma^T(x(k+1)) D x(k+1) - \sigma^T(x(k)) D x(k) \\
&\quad - \sum_{i=1}^{n} d_i \int_{x_i(k)}^{x_i(k+1)} \sigma(s) ds \\
&= -\frac{1}{2} z^T(k+1) D W z(k+1) \\
&\quad - z^T(k+1) D x(k+1) \\
&\quad + \sigma^T(x(k+1)) D x(k+1) - \sigma^T(x(k)) D x(k) \\
&\quad - \sum_{i=1}^{n} d_i \int_{x_i(k)}^{x_i(k+1)} \sigma(s) ds \\
&= -\frac{1}{2} z^T(k+1) D W z(k+1) \\
&\quad + \sigma^T(x(k)) D [x(k+1) - x(k)] \\
&\quad - \sum_{i=1}^{n} d_i \int_{x_i(k)}^{x_i(k+1)} \sigma(s) ds \\
&= -\frac{1}{2} z^T(k+1) D W z(k+1) \\
&\quad - \sum_{i=1}^{n} d_i \int_{x_i(k)}^{x_i(k+1)} \left[ \sigma(s) - \sigma(x_i(k)) \right] ds
\end{aligned}$$

for all $k \geq 0$.

Using Lemma 8.3,

$$\int_{x_i(k)}^{x_i(k+1)} \left[ \sigma(s) - \sigma(x_i(k)) \right] ds \geq \frac{\left[ \sigma(x_i(k+1)) - \sigma(x_i(k)) \right]^2}{2}$$

for $k \geq 0$ and $(i = 1, \cdots, n)$.

Then,

$$\begin{aligned}
E(x(k+1)) - E(x(k)) &\leq -\frac{1}{2} z^T(k+1) D \left( I + W \right) z(k+1) \\
&\leq -\lambda \parallel \sigma(x(k+1)) - \sigma(x(k)) \parallel^2
\end{aligned}$$

for all $k \geq 0$, where $\lambda > 0$ is the smallest eigenvalue of the matrix $I + W$. That is

$$\parallel \sigma(x(k+1)) - \sigma(x(k)) \parallel^2 \leq \frac{1}{\lambda} \left[ E(x(k)) - E(x(k+1)) \right] \qquad (8.8)$$

Clearly, $E(x(k))$ is monotone decrease. Since the network (8.2) is bounded, then $E(k)$ is bounded also. Hence, the sequence $\{E(k)\}$ is a monotone bounded sequence. A limit must exist. That is, there exists a constant $E_0$ such that

$$\lim_{k \to +\infty} E(x(k)) = E_0.$$

By (8.8), it follows that

$$\sum_{k=0}^{\infty} \| \sigma(x(k+1)) - \sigma(x(k)) \|^2 \leq \frac{1}{\lambda} [E(x(0)) - E_0] < +\infty. \tag{8.9}$$

Thus,

$$\lim_{k \to +\infty} \| \sigma(x(k+1)) - \sigma(x(k)) \| = 0. \tag{8.10}$$

Using Lemma 8.4, each component of $x(k+1)$ is a polynomial with order no more than $k$, it follows that $\| \sigma(x(k+1)) - \sigma(x(k)) \|^2$ is also a polynomial with order no more than $k$. Together with (8.9) and (8.10), there must exist constants $M > 0$ and $\eta > 0$ such that

$$\| \sigma(x(k+1)) - \sigma(x(k)) \| \leq Me^{-\eta k}$$

for $k \geq 0$.

Given any $\epsilon > 0$, there exist a $K > 0$ such that

$$\frac{Me^{-\eta K}}{1 - e^{-\eta}} \leq \epsilon.$$

Given any $k_1 > k_2 \geq K$, it follows that

$$\begin{aligned}
& \|\sigma(x(k_1)) - \sigma(x(k_2))\| \\
= & \left\| \sum_{r=k_2}^{k_1 - 1} \left[\sigma(x(r+1)) - \sigma(x(r))\right] \right\| \\
\leq & \sum_{r=k_2}^{k_1 - 1} \|\sigma(x(r+1)) - \sigma(x(r))\| \\
\leq & M \cdot \sum_{r=k_2}^{k_1 - 1} e^{-\eta r} \\
\leq & Me^{-\eta K} \cdot \sum_{r=0}^{+\infty} \left(e^{-\eta}\right)^r \\
\leq & \frac{Me^{-\eta K}}{1 - e^{-\eta}} \\
\leq & \epsilon.
\end{aligned}$$

This shows that the sequence $\{\sigma(x(k))\}$ is a *Cauchy sequence*. By *Cauchy Convergence Principle*, there must exist a constant $z^*$ such that

$$\lim_{k\to+\infty} \sigma(x(k)) = z^*.$$

Then, from (8.2),

$$\lim_{k\to+\infty} x(k+1) = W \cdot \lim_{k\to+\infty} \sigma(x(k)) + h = Wz^* + h \triangleq x^e.$$

Clearly, $\sigma(x^e) = z^*$, and $x^e$ is an equilibrium point of (8.2). This completes the proof.

The condition that $D(I+W)$ is positive definite in the Theorem 8.2 is critical. Consider for an example, a one dimensional network

$$x(k+1) = -\sigma(x(k)) + 1$$

for $k \geq 0$. Since it is one dimensional, choose $D = 1$, then $D(I+W) = 0$ which is not positive definite. Clearly, the trajectory starting from zero does not converge.

It could be seen that the results of Theorem 8.2 could be generalized to more general discrete neural networks.

## 2.4 Simulation Examples

In this section, some examples and simulation results to illustrate and verify the theory developed will be provided.

EXAMPLE 8.1 *Consider the two-dimensional neural network:*

$$\begin{bmatrix} x_1(k+1) \\ x_2(k+1) \end{bmatrix} = \begin{bmatrix} 0 & -2 \\ -2 & 0 \end{bmatrix} \begin{bmatrix} \sigma(x_1(k)) \\ \sigma(x_2(k)) \end{bmatrix} + \begin{bmatrix} 3 \\ 3 \end{bmatrix} \tag{8.11}$$

*for* $k \geq 0$.

Clearly, the network satisfies the conditions of Theorem 8.1, thus it is bounded and there exists a compact set to attract all the trajectories of the network. By simple calculations, it gives $a_1 = a_2 = 0$, $b_1 = b_2 = 3$. Thus, we have an attracting set $S = \{x | 0 \leq x_1, x_2 \leq 3\}$.

It is easy to see that this network has three equilibrium points $(-3, 3)$, $(1, 1)$ and $(3, -3)$. They are all located in $S$. In addition, this network has many periodic trajectories. For examples,

$$\left\{ \begin{bmatrix} 3 \\ 3 \end{bmatrix}, \begin{bmatrix} -3 \\ -3 \end{bmatrix} \right\}; \quad \left\{ \begin{bmatrix} 1 \\ 3 \end{bmatrix}, \begin{bmatrix} -3 \\ 1 \end{bmatrix} \right\}$$

are two periodic trajectories. All the periodic trajectories of (9) are also located in $S$. Each trajectory of (8.1) will go into $S$ and remains in $S$ indefinitely.

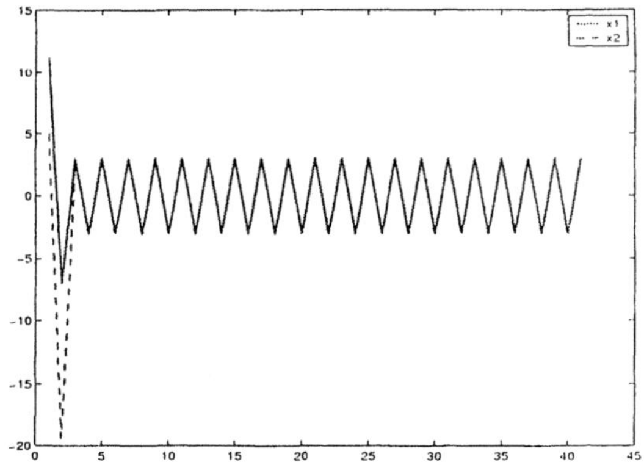


*Figure 8.1.* Periodic trajectory of network (8.1)

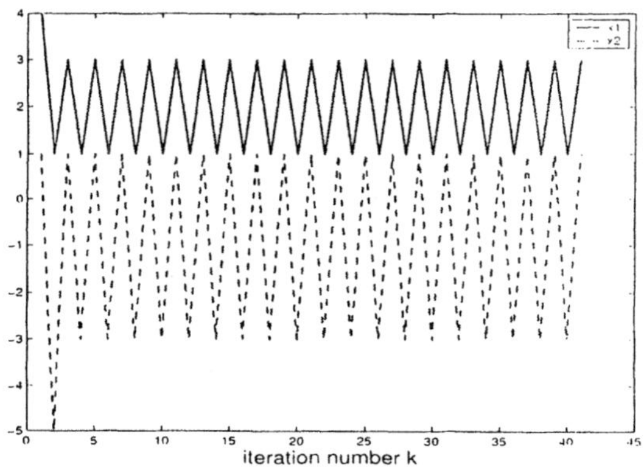


*Figure 8.2.* Periodic trajectory of network (8.1)

Figure 8.1 and Figure 8.2 show two trajectories of (8.1) enter $S$ and converge to periodic trajectories.

EXAMPLE 8.2 *Consider the two-dimensional neural network*

$$\begin{bmatrix} x_1(k+1) \\ x_2(k+1) \end{bmatrix} = \begin{bmatrix} 0.6 & -0.2 \\ -0.4 & 0.8 \end{bmatrix} \begin{bmatrix} \sigma(x_1(k)) \\ \sigma(x_2(k)) \end{bmatrix} + \begin{bmatrix} 1 \\ 1 \end{bmatrix} \tag{8.12}$$

*for* $k \geq 0$.

It is easy to see that the conditions of Theorem 8.1 are satisfied. Using Theorem 8.1, the network is bounded and there exists a compact set to attract the network. By simple calculations, it gives

$$S = \left\{ \begin{bmatrix} x_1 \\ x_2 \end{bmatrix} \middle| -1 \leq x_1 \leq 5, -2 \leq x_2 \leq 5 \right\}$$

which globally attracts the network.

Define a matrix

$$D = \begin{bmatrix} 2 & 0 \\ 0 & 1 \end{bmatrix}.$$

Then, we have

$$DW = \begin{bmatrix} 1.2 & -0.4 \\ -0.4 & 0.8 \end{bmatrix}$$

which is a symmetric positive definite matrix. Thus, the conditions of Theorem 8.2 are satisfied. By Theorem 8.2, the network is completely convergent.

All the equilibrium points are located in the attractive set $S$. The set of all equilibrium points can be calculated out as

$$S^e = \left\{ \begin{bmatrix} x_1 \\ x_2 \end{bmatrix} \middle| 2x_1 + x_2 + 5 = 0, x_1 \geq 0, x_2 \geq 0 \right\}.$$

Clearly, $S^e$ is a linear attractor of the network (8.12). This network has a lot of equilibrium points and the equilibrium points are not isolated.

Figure 8.3 shows the simulation results for complete stability and global attractivity of the network (8.12) for 250 trajectories originating from randomly selected initial points. It can be observed that every trajectory converges to the equilibrium set. Clearly, this network has many equilibrium points and the equilibrium points are not isolated. The square contained part in Figure 4 is the global attracting compact set $S$. All of the equilibrium points are located within this compact set $S$.

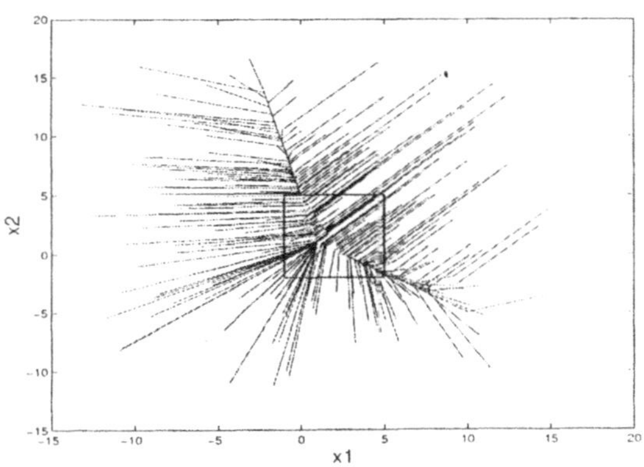


*Figure 8.3.* Global attractivity and complete stability of network (8.12)

EXAMPLE 8.3 *Consider the following three-dimensional neural network:*

$$x(k+1) = W\sigma(x(k)) + h, \tag{8.13}$$

*for* $k \geq 0$, *where* $x = (x_1, x_2, x_3)^T$, $h = (3, -3, 3)^T$ *and*

$$W = \begin{bmatrix} 0 & -3 & 0.5 \\ -7 & 0.5 & -4 \\ 0.5 & -2 & 0 \end{bmatrix}.$$

This network satisfies the conditions of Theorem 8.1. Thus, it is bounded and there exists a compact set to attract all the trajectories. With simple calculations, it follows

$$S = \{-18 \leq x_1 \leq 6, -69 \leq x_2 \leq 6, -12 \leq x_3 \leq 6\}.$$

Figure 8.4 shows the simulation results in three-dimensional space for 50 trajectories originating from randomly selected initial points.

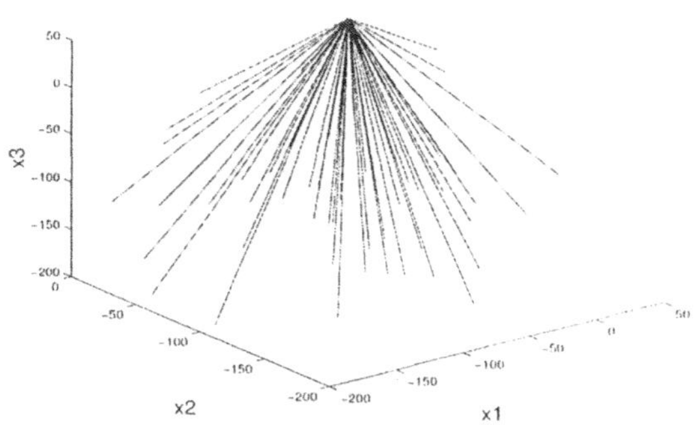


*Figure 8.4.* Non-divergence and global attractivity of network (8.13)

From the simulation results, all the trajectories in Figure 8.4 converge to a point $(6, -69, 6)^T$. It seems that this network has the property of global convergence. However, we cannot prove this property at present. It is noted that the spectral radius of the matrix $W$ is larger than 1, which differs from the usual requirement to be less than 1 for global stability of discrete systems. To the best of our knowledge, at the moment, there is no theory which can show whether this network is globally stable or not.

Apart from the presented simulations, we have extensively simulated the non-divergence, global attractivity and complete stability of networks that belong to the network model (8.2). All the results verify the theory developed.

## 3. Discrete RNN Model for Winner Take All Problem

The problem of winner-take-all is to select the largest element from a data set. Using neural networks to winner-take-all problem has been studied widely by many authors in recent years, see for examples [2, 16, 38, 45, 48, 62, 75, 82, 98, 141, 162, 178, 189, 197, 198, 227]. It is well known that they play an important role in the design for many unsupervised learning networks. Winner-take-all networks for selecting the most prominent element are central processing components in competitive learning neural networks and nearest neighbor pattern classifiers [82, 189, 197, 198].

So far, many kinds of neural network models have been proposed to implement the winner-take-all function. In [48], based on Hopfield neural network topology, an alternative neural network is designed as winner-take-all function. Winner-take-all cellular neural networks were firstly studied in [162]. It was shown that cellular neural networks can perform winner-take-all function if parameters of the networks satisfy some conditions. However, the sizes of the networks are limited to less than 10 neurons. [2] overcame this problem and improved the robustness of [162] by introducing a clipped total feedback. In [62] and [16], Lotka-Volterra competitive neural networks have been used to implement winner-take-all, winner-share-all and $k$-winner-take-all functions. Another $k$-winner-take-all analog neural network was reported in [28]. Recently, [178] proposed a simple winner-take-all neural network model with fewer parameters. The network response time was deeply studied in [178].

Many of the existing winner-take-all models are formulated by continuous neural networks. In this section, a discrete recurrent neural networks model to implement winner-take-all behaviour will be proposed. The proposed neural network model has simple organizations and well understood dynamic behaviours.

## 3.1 Neural Network Model

The proposed winner-take-all discrete recurrent neural network model is as follows

$$\begin{cases} x_i(k+1) = \dfrac{(c+\beta)(x_i(k)+\gamma)^{\alpha}}{\sum\limits_{j=1}^{n}(x_j(k)+\gamma)^{\alpha}} - \gamma \\ y_i(k) = h(x_i(k)), (i = 1, \cdots, n) \end{cases} \tag{8.14}$$

where $c \geq 1, \alpha > 1, \beta > \gamma > 0, y_i(t)(i = 1, \cdots, n)$ are output of the network, $h(s)$ is a piecewise linear function defined as

$$h(s) = \begin{cases} c, & \text{if } s > c \\ s, & \text{if } 0 \leq s \leq c \\ 0, & \text{if } s < 0. \end{cases}$$

See Figure 8.5 for the intuitive explanation of the output function. Figure 8.6 shows the structure of the network model. It should be noted that the output function is not involved in the state equation of the network. It is used to detect the state of the network to see whether a neuron reaches zero or one.

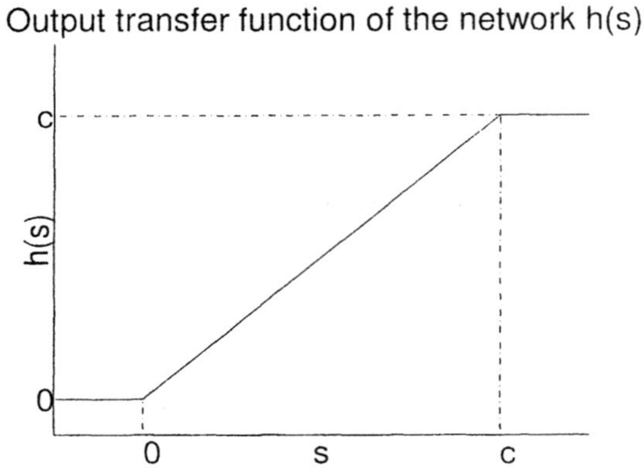


*Figure 8.5.* Output transfer function of the network $h(s)$.

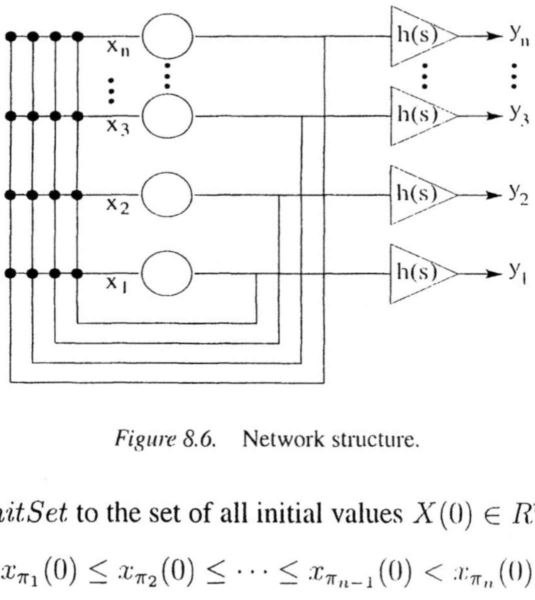


*Figure 8.6.* Network structure.

Denote by $InitSet$ to the set of all initial values $X(0) \in R^n$ that satisfies

$$0 < x_{\pi_1}(0) \leq x_{\pi_2}(0) \leq \cdots \leq x_{\pi_{n-1}}(0) < x_{\pi_n}(0) < c$$

where each $\pi_i \in \{1, \cdots, n\}$.

The network is said to has the winner-take-all behaviour, if for any $X(0) \in InitSet$ there exist $K_{\pi_1} \leq K_{\pi_2} \leq \cdots \leq K_{\pi_n}$ such that $y_{\pi_n}(k) = c$ for all $k \geq K_{\pi_n}$ and $y_{\pi_i}(k) = 0 (i = 1, \cdots, n-1)$ for all $k \geq K_{\pi_i}$. The $\pi_n^{th}$ neuron is called the winner and all other neurons are called losers.

## 3.2 Dynamic Properties

In this section, the dynamical properties of the proposed winner-take-all discrete recurrent neural network will be studied. It will show that the network has clear dynamic behaviours. The first theorem will get the representation of the solutions of the state equation of the network.

THEOREM 8.3 *For any $X(0) \in InitSet$, the solution of (8.14) starting from $X(0)$ can be represented as*

$$x_i(k+1) = \frac{c+\beta}{\sum_{j=1}^{n} \left(\frac{x_j(0)+\gamma}{x_i(0)+\gamma}\right)^{\alpha^k}} - \gamma \tag{8.15}$$

*for all $k \geq 0$ and $i = 1, \cdots, n$.*

*Proof*: Since $X(0) \in InitSet$, thus $x_i(0) + \gamma > 0, (i = 1, \cdots, n)$. It is easy to see from (8.14) that $x_i(k) + \gamma > 0 (i = 1, \cdots, n)$ for all $k \geq 0$.

From (8.14) it follows that

$$\frac{x_j(k+1)+\gamma}{x_i(k+1)+\gamma} = \left(\frac{x_j(k)+\gamma}{x_i(k)+\gamma}\right)^{\alpha}$$

and so

$$\frac{x_j(k+1)+\gamma}{x_i(k+1)+\gamma} = \left(\frac{x_j(0)+\gamma}{x_i(0)+\gamma}\right)^{\alpha^{k+1}}$$

for all $k \geq 0$. Then,

$$\begin{aligned} x_i(k+1) &= \frac{(c+\beta)(x_i(k)+\gamma)^{\alpha}}{\sum_{j=1}^{n}(x_j(k)+\gamma)^{\alpha}} - \gamma \\ &= \frac{c+\beta}{\sum_{j=1}^{n}\left(\frac{x_j(k)+\gamma}{x_i(k)+\gamma}\right)^{\alpha}} - \gamma \\ &= \frac{c+\beta}{\sum_{j=1}^{n}\left(\frac{x_j(0)+\gamma}{x_i(0)+\gamma}\right)^{\alpha^k}} - \gamma \end{aligned}$$

for all $k \geq 0$. The proof is completed.

THEOREM 8.4 *Suppose that $X(0) \in InitSet$. If $x_i(0) < x_j(0)$ then $x_i(k) < x_j(k)$ for all $k \geq 0$. If $x_i(0) = x_j(0)$ then $x_i(k) = x_j(k)$ for all $k \geq 0$.*

*Proof*: It is easy to see from (8.14) that

$$\frac{x_i(k)+\gamma}{x_j(k)+\gamma} = \left(\frac{x_i(0)+\gamma}{x_j(0)+\gamma}\right)^{\alpha^k}$$

for all $k \geq 0$. If $x_i(0) < x_j(0)$ then

$$\frac{x_i(0)+\gamma}{x_j(0)+\gamma} < 1.$$

and so

$$\frac{x_i(k)+\gamma}{x_j(k)+\gamma} < 1.$$

That is $x_i(k) < x_j(k)$ for all $k \geq 0$. The same argument can be applied to the case of $x_i(0) = x_j(0)$. The proof is completed.

Theorem 8.4 shows that the states of any two different neurons in the network will never intersect each other if their corresponding initial values are different, and if their corresponding initial values are equal then their states will remain identical.

THEOREM 8.5 *For any $X(0) \in InitSet$, if $x_m(0) > x_j(0)(j \neq m; j = 1, \cdots, n)$ then*

$$x_i(k) \to \begin{cases} c+\beta-\gamma & \text{if } i = m \\ -\gamma & \text{otherwise} \end{cases}$$

*as $k \to +\infty$.*

*Proof:* Since $x_m(0) > x_j(0)(j \neq m; j = 1, \cdots, n)$, then

$$\left(\frac{x_j(0)+\gamma}{x_m(0)+\gamma}\right)^{\alpha^k} \to 0 (i \neq m; j = 1, \cdots, n)$$

as $k \to +\infty$. Hence, it follows from (8.15) that

$$\begin{aligned} x_m(k+1) &= \frac{c+\beta}{1+\sum\limits_{j\neq m, j=1}^{n}\left(\frac{x_j(0)+\gamma}{x_m(0)+\gamma}\right)^{\alpha^k}} - \gamma \\ &\to c+\beta-\gamma \end{aligned}$$

as $k \to +\infty$.

For any $i \neq m$,

$$\left(\frac{x_m(0)+\gamma}{x_i(0)+\gamma}\right)^{\alpha^k} \to +\infty$$

as $k \to +\infty$, and so by (8.15) that

$$\begin{aligned} x_i(k+1) &= \frac{c+\beta}{\left(\frac{x_m(0)+\gamma}{x_i(0)+\gamma}\right)^{\alpha^k} + \sum\limits_{j\neq m, j=1}^{n}\left(\frac{x_j(0)+\gamma}{x_i(0)+\gamma}\right)^{\alpha^k}} - \gamma \\ &\to -\gamma \end{aligned}$$

as $k \to +\infty$. The proof is completed.

COROLLARY 8.1 *For any $X(0) \in InitSet$, if $x_m(0) > x_j(0)(j \neq m; j = 1, \cdots, n)$ then there exists a $k_m > 0$ such that $y_m(k) = c$ for all $k \geq k_m$. And there exists a $k_i > 0$ such that $y_i(k) = 0$ for all $k \geq k_i$.*

Theorem 8.5 and the Corollary 8.1 together show the convergence of the state of the network. It is shown that only the state of the winner will approach $c+\beta-\gamma$ and all the losers of their states will approach $-\gamma$. Hence, the output of the winner will reach $c$ in finite iteration steps and the output of each loser will reach zero in finite iteration steps. It should be noted that $\gamma > 0$ and $\beta > \gamma$ are important conditions, otherwise both the losers and the winner will not reach zero and $c$ respectively in finite iteration steps.

THEOREM 8.6 *Suppose that $\frac{\beta-c}{2} < \gamma$, then for any $X(0) \in InitSet$, if $x_m(0) > x_j(0)(j \neq m; j = 1, \cdots, n)$ we have $x_i(k) < c(i \neq m; i = 1, \cdots, n)$ for all $k \geq 0$.*

*Proof*: Since $\frac{\beta-c}{2} < \gamma$, then for any $i \neq m$, it follows from (8.15) that

$$\begin{aligned} x_i(k+1) &= \frac{c+\beta}{\sum_{j=1}^{n} \left( \frac{x_j(0)+\gamma}{x_i(0)+\gamma} \right)^{\alpha^k}} - \gamma \\ &< \frac{c+\beta}{1+\left( \frac{x_m(0)+\gamma}{x_i(0)+\gamma} \right)} - \gamma \\ &< \frac{c+\beta}{2} - \gamma \\ &< c. \end{aligned}$$

This completes the proof.

Theorem 8.6 provides conditions to guarantee the output of each loser will never reach $c$.

THEOREM 8.7 *Suppose that*

$$\gamma < \frac{c+\beta}{n}$$

*then for any $X(0) \in InitSet$, if $x_m(0) > x_j(0)(j \neq m; j = 1, \cdots, n)$ we have $x_m(k) > 0$ for all $k \geq 0$.*

*Proof*: If $x_m(0) > x_j(0)(j \neq m; j = 1, \cdots, n)$, then

$$\frac{x_j(0)+\gamma}{x_m(0)+\gamma} < 1, (j \neq m; j = 1, \cdots, n)$$

and so from (8.15),

$$\begin{aligned} x_m(k+1) &= \frac{c+\beta}{\sum_{j=1}^{n}\left(\frac{x_j(0)+\gamma}{x_m(0)+\gamma}\right)^{\alpha^k}} - \gamma \\ &> \frac{c+\beta}{n} - \gamma \\ &> 0. \end{aligned}$$

This completes the proof.

Theorem 8.7 provides conditions for the output of the winner will never reach zero.

Next, the conditions under which the winner neuron is the last one settling will be discussed.

THEOREM 8.8 *Suppose $\beta - \gamma < \gamma$, then for any $X(0) \in InitSet$ if $x_m(0) > x_s(0) \geq x_j(0) (j \neq m; j \neq s; j = 1, \cdots, n)$, $y_m(k) = c$ implies $y_s(k) = 0$.*

*Proof*: It is easy to see from (8.14) that

$$\sum_{i=1}^{n}[x_i(k)+\gamma] = c + \beta$$

for all $k \geq 1$. Then

$$x_m(k) = c + \beta - \gamma - \sum_{i \neq m, i=1}^{n} (x_i(k)+\gamma)$$

If $y_m(k) = c$, then $x_m(k) \geq c$ and so

$$c + \beta - \gamma - \sum_{i \neq m, i=1}^{n} (x_i(k)+\gamma) > c.$$

Hence, we have $\beta - \gamma > x_s(k) + \gamma$. This implies that $x_s(k) < 0$. That is $y_s(k) = 0$. The proof is completed.

## 3.3 Network Response Time

For any $X(0) \in InitSet$, suppose $c > x_m(0) > x_s(0) \geq x_j(0) > 0 (j \neq m; j \neq s; j = 1, \cdots, n)$. Then $x_s(0)$ is one of the second largest initial state value. From the last section, it is known that $y_s(k)$ will reach and then remains zero after some large $k$. The smallest $k$ such that $y_s(k) = 0$ is called the response time of the network and denote it by $k_{resp}$.

For the network model of (8.14), it seems difficult to calculate the exact response time. Next, an upper bound for the response time of the network will be calculated. From (8.15), it follows that

$$x_s(k+1) < \frac{c+\beta}{\left(\frac{x_m(0)+\gamma}{x_s(0)+\gamma}\right)^{\alpha^k}} - \gamma$$

then if

$$\frac{c+\beta}{\left(\frac{x_m(0)+\gamma}{x_s(0)+\gamma}\right)^{\alpha^k}} - \gamma < 0$$

that is

$$k > \left\lceil \frac{\ln\left(\frac{\ln\left(\frac{c+\beta}{\gamma}\right)}{\ln\left(\frac{x_m(0)+\gamma}{x_s(0)+\gamma}\right)}\right)}{\ln \alpha} \right\rceil = K$$

then $x_s(k+1) < 0$. Hence, the above $K$ is an upper bound of the response time. That is $k_{resp} \leq K$. The $K$ can be used to estimate the response time of the network. In the next section, simulation results will show that this upper bound of response time is very close to the exact response time.

## 3.4 Simulations

The dynamic properties in last subsection clearly shows that the network of (8.14) performs the winner-take-all function if the initial values and parameters of the network satisfy some very simple conditions as described follows.

For each input initial value $X(0) = (x_1(0), \cdots, x_n(0))^T \in R^n$, suppose that that $0 < x_i(0) < c (i = 1, \cdots, n)$ and there is only one largest component of $X(0)$. This means that there is a unique $m \in \{1, \cdots, n\}$ such that $x_m(0) > x_j(0) (j \neq m; j = 1, \cdots, n)$.

For the parameters of the network, require that

$$\begin{cases} \alpha > 1 \\ \beta > 0 \\ \max\{\frac{\beta - c}{2}, \frac{\beta}{2}\} < \gamma < \min\{\beta, \frac{c+\beta}{n}\} \end{cases}$$

These conditions are quite easy to be implemented.

In the above conditions, $\alpha > 1$ is used to control the converge rate of the network. The larger the $\alpha$ is, the less iteration steps will be. However, for large $\alpha$ the computation will become more complex and will take more time. So, this is a trade off. It is difficult to get balance between $\alpha$ and the computational complexity. $\beta > \gamma$ is used to guarantee the winner will go across the bound of $c$ so that the output of winner will reach $c$ in finite iteration steps. Similarly,

$\gamma > 0$ is used to guarantee the output of the losers reach zero in finite iteration steps. $\frac{\beta - c}{2} < \gamma$ guarantees that the output of each loser will never reach $c$ and $\gamma < \frac{c+\beta}{n}$ guarantees the output of the winner will never reach zero. The condition $\gamma > \frac{\beta}{2}$ guarantees the winner is the last one settling.

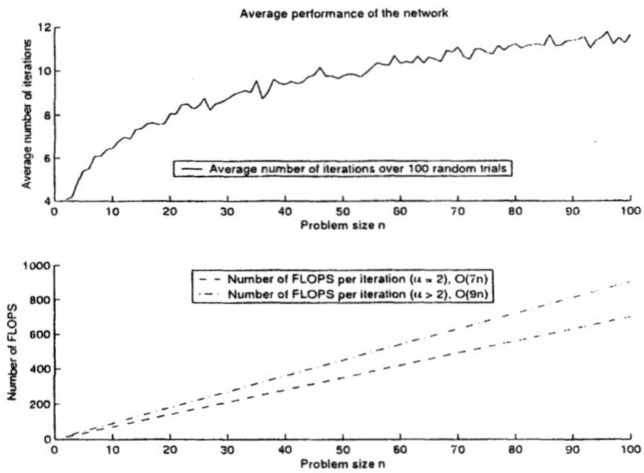


*Figure 8.7.* Comprehensive study on the performance of the network.

Upper part of Figure 8.7 shows the average number of iterations taken for different problem size $n$. There are 100 random testing cases for each problem size (number of neurons). There is a leveling off trend in the curve. Lower part of the figure gives two straight lines describing the number of floating point operations (FLOPS) involved in each iteration for different problem size. When $\alpha = 2$, less computation is required since the squaring operations in the model could be done by multiplications. However, when $\alpha > 2$, more computations are needed to perform the exponentiation. Nevertheless, the number of FLOPS required per iteration, no matter how $\alpha$ is set, is in the order of $O(n)$.

The left column of Figure 8.8 shows three scenarios of invoking the network with different $\alpha$'s. The right column shows how the network performs in different scale of problems. All experiments agree that the model works accurately and efficiently provided that the initial conditions and network parameters are given according to our simple constraints.

The above simulation shows that under some simple conditions for the initial values and the parameters of the network, the network acts the winner-take-all function correctly with satisfactory efficiency.

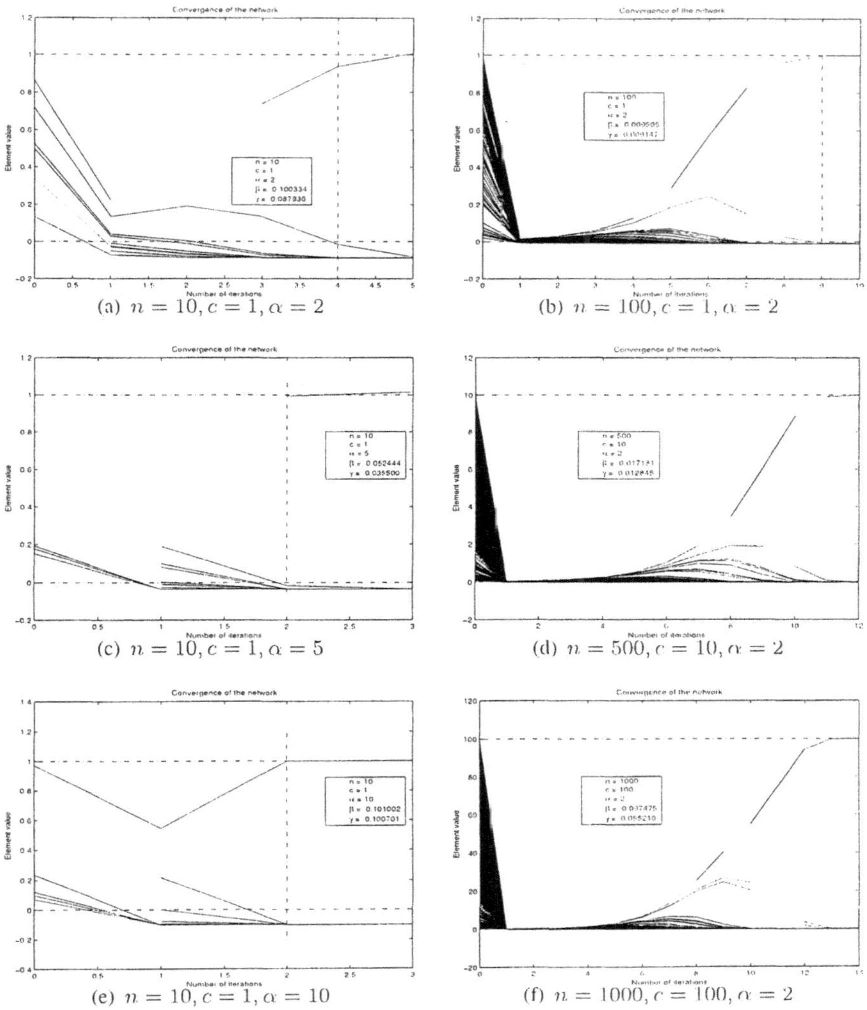


(a) $n = 10, c = 1, \alpha = 2$

(b) $n = 100, c = 1, \alpha = 2$

(c) $n = 10, c = 1, \alpha = 5$

(d) $n = 500, c = 10, \alpha = 2$

(e) $n = 10, c = 1, \alpha = 10$

(f) $n = 1000, c = 100, \alpha = 2$

*Figure 8.8.* Simulation results of various parameters. The vertical dotted line shows the theoretical upper bound on the number of iterations required to have all other losers drop below zero.

# References

[1] S. Abe, Global convergence and suppression of spurious states of the Hopfield neural networks, *IEEE Trans. Circuits Syst. I*, vol. 40, no. 4, pp. 246-257, 1993.

[2] L. Andrew, Improving the robustness of winner-take-all cellular neural networks, *IEEE Trans. Circuits Syst. II*, vol. 43, no. 4, pp.329-334, 1996.

[3] D. Angeli, E. Sontag and Y. Wang, A characterization of integral input-to-state stability, *IEEE Trans. Automatic Control*, vol. 45, no. 6, pp. 1082-1097, 2000.

[4] N. Ansari, E. S. H. Hou and Y. Yu, A new method to optimize the satellite broadcasting schedules using the mean field annealing of a Hopfield neural network, *IEEE Trans. Neural Networks*, vol. 6, no. 2, pp. 470-483, 1995.

[5] P. Arena, S. Baglio and L. Fortuna, Chua's circuit can be generated by CNN cells, *IEEE Trans. Circuits Syst. I*, vol. 42, no. 2, pp. 123-125, 1995.

[6] P. Arena, S. Baglio, L. Fortuna and G. Manganaro, How state controlled CNN cells generate the dynamics of the Colpitts-like oscillator, *IEEE Trans. Circuits Syst. I*, vol. 43, no. 7, pp. 602-605, 1996.

[7] P. Arena, R. Caponetto, L. Fortuna and A. Rizzo, Noise-supported wavefronts in cellular neural networks based circuits, *IEEE Trans. Circuits Syst. I*, vol. 48, no. 3, pp. 360-363, 2001.

[8] S. Arik, Stability analysis of delayed neural networks, *IEEE Trans. Circuits Syst. I*, vol. 47, no. 7, pp.1089-1092, 2000.

[9] S. Arik, An analysis of global asymptotic stability of delayed cellular neural networks, *IEEE Trans. Neural Networks*, vol. 13, no. 5, pp. 1239-1242, 2002.

[10] S. Arik and V. Tavsanoglu, Equilibrium analysis of delayed CNN's, *IEEE Trans. Circuits Syst. I*, vol. 45, no. 2, pp. 168-171, 1998.

[11] S. Arik and V. Tavsanoglu, A comment on 'Necessary and sufficient condition for absolute stability of neural networks', *IEEE Trans. Circuits Syst. I*, vol. 45, no. 5, pp. 595-596, 1998.

[12] S. Arik and V. Tavsanoglu, On the global asymptotic stability of delayed cellular neural networks, *IEEE Trans. Circuits Syst. I*, vol. 47, no. 4, pp. 571-574, 2000.

[13] S. Arik and V. Tavsanoglu, Equilibrium analysis of nonsymmetric CNN's, *Int. J. Circuit Theory Appl.*, vol. 24, pp. 269-274, 1996.

[14] S. Arik and V. Tavsanoglu, A sufficient conditions for absolute stability of a large class of dynamical neural networks, *IEEE Trans. Circuits Syst. I*, vol. 47, no. 5, pp. 758-760, 2000.

[15] T. Asai, T. Fukai and S. Tanaka, A subthreshold MOS circuit for the Lotka-Volterra neural network producing the winner-share-all solution, *Neural Networks*, vol. 12, pp. 211-216, 1999.

[16] T. Asai, M. Ohtani and H. Yonezu, Analog integrated circuits for the Lotka-Volterra competitive neural networks, *IEEE Trans. Neural Networks*, vol. 10, no. 5, pp. 1222-1231, 1999.

[17] P. Baldi and A. F. Atiya, How delays affect neural dynamics and learning, *IEEE Trans. Neural Networks*, vol. 5, no. 4, pp.612-621, 1994.

[18] S. H. Bang, B. J. Sheu and T. H. Y. Sheu, Optimal solutions for cellular neural networks by paralleled hardware annealing, *IEEE Trans. Neural Networks*, vol. 7, no. 2, pp. 440-454, 1996.

[19] R. Ben-Yishai, R. Lev Bar-Or and H. Sompolinsky, Theory of orientation tuning in visual cortex, *Proc. Natl. Acad. Sci. USA*, vol. 92, pp. 3844-3848, 1995.

[20] A. Berman and R. J. Plemmons, Nonnegative matrices in the mathematical science, New York: Academic Press, 1994.

[21] S. Bharitkar and J. M. Mendel, The hysteretic Hopfield neural network, *IEEE Trans. Neural Networks*, vol. 11, no. 4, pp. 879-888, 2000.

[22] S. Bharitkar, K. Tsuchiya and Y. Takefuji, Microcode optimization with neural networks, *IEEE Trans. Neural Networks*, vol. 10, no. 3, pp. 698-703, 1999.

[23] A. Bhaya, E. Kaszkurewicz, and V. S. Kozyakin, Existence and stability of a unique equilibrium in continuous-valued discrete-time asynchronous Hopfield neural networks, *IEEE Trans. Neural Networks*, vol. 7, pp. 620–628, May 1996.

[24] M. Biey, M. Gilli and P. Checco, Complex dynamic phenomena in space-invariant cellular neural networks, *IEEE Trans. Circuits Syst. I*, vol. 49, no. 3, pp. 340-345, 2002.

[25] A. Bouzerdoum and T. R. Pattison, Neural network for quadratic opti-mization with bound constraints, *IEEE Trans. Neural Networks*, vol. 4, pp. 293–303, 1993.

[26] R. K. Brouwer, Automatic Growing of a Hopfield Style Network During Training for Classification, vol. 10, no. 3, pp. 529-537, 1997.

[27] M. Brucoli, L. Carnimeo, and G. Grassi, Discrete-time cellular neural networks for associative memories with learning and forgetting capabilities, *IEEE Trans. Circ. Syst. I*, vol. 42, pp. 396–399, July, 1995.

[28] B. D. Calvert and C. A. Marinov, Another K-winners-take-all analog neural network, *IEEE Trans. Neural Networks*, Vol. 11 no. 4 , pp. 829 -838, 2000.

[29] J. Cao and D. Zhou, Stability analysis of delayed cellular neural networks, *Neural Networks*, vol. 11, pp.1601-1605, 1998.

[30] J. Cao and L. Wang, Exponential stability and periodic oscillatory solution in BAM networks with delays, *IEEE Trans. Neural Networks*, vol. 13, no. 2, pp. 457 -463, 2002.

[31] C. K. Chan and L. M. Cheng, The convergence properties of a clipped Hopfield network and its application in the design of keystream generator, *IEEE Trans. Neural Networks*, vol. 12, no. 2, pp. 340-348, 2001.

[32] Y. Chen, Global stability of nerual networks with distributed delays, *Neural Networks*, vol. 15, pp. 867-871, 2002.

[33] T. Chen, Global exponential stability of delayed Hopfield neural networks, *Neural Networks*, vol. 14, no. 8, pp. 977-980, 2001.

[34] T. Chen and S. I. Amari, Exponential convergence of delayed dynamical systems, *Neural Computations*, vol. 13, pp. 621-635, 2001.

[35] T. Chen and S. I. Amari, New theorems on global convergence of some dynamical systems, *Neural Neworks*, vol. 14, no. 3, pp. 251-255, 2001.

[36] T. Chen and S. I. Amari, Stability of asymmetric Hopfield networks, *IEEE Trans. Neural Networks*, vol. 12 no. 1, pp: 159-163, 2001.

[37] Y. H. Chen and S. C. Fang, Neurocomputing with time delay analysis for solving convex quadratic programming problems, *IEEE Tran. Neural Networks*, vol. 11, no. 1, pp. 230-240, 2000.

[38] K. S. Cheng, J. S. Lin and C. W. Mao, The application of competitive Hopfield neural network to medical image segmentation, *IEEE Trans. Medical Imaging*, vol. 15 no. 4, pp.560-567, 1996.

[39] L. O. Chua and T. Roska, Stability of a class of nonreciprocal cellular neural networks, *IEEE Trans. Curcuits Syst.*, vol. 37, pp. 1520-1527, Dec. 1990.

[40] L. O. Chua and T. Roska, The CNN paradigm, *IEEE Trans. Circuits Syst. I*, vol . 40, no. 3, pp. 147-156, 1993.

[41] L. O. Chua and C. W. Wu, On the universe of stable cellular neural networks, *Int. J. Circuits Theory Appl.*, vol. 20, pp. 497-517, July-Aug. 1992.

[42] L. O. Chua and L. Yang, Cellular neural networks: Theory, *IEEE Trans. Circuits Syst.*, vol. 35, pp.1257-1272, 1988.

[43] L. O. Chua and L. Yang, Cellular neural networks: Applications, *IEEE Trans. Circuits Syst.*, vol. 35, pp. 1273-1290, 1988.

[44] P. P. Civalleri, M. Gilli and L. Pandolfi, On the stability of cellular neural networks with delay, *IEEE Trans. Circuits Syst. I*, vol. 40, no. 3, pp.157-165, 1993.

[45] M. A. Cohen and S. Grossberg, Absolute stability of global pattern formation and parallel memory storage by competitive neural networks, *IEEE Trans. Systems, Man, Cybernetics*, vol. 13, pp. 815-826, 1983.

[46] K. R. Crounse and L. O. Chua, Methods for image processing and pattern formation in Cellular Neural Networks: a tutorial, *IEEE Trans. Circuits Syst. I*, vol. 42, no. 10, pp. 583-601, 1995.

[47] J. M. Cruz and L. O. Chua, Application of cellular neural networks to model population dynamics, *IEEE Trans. Circuits Syst. I*, vol. 42, no. 10, pp. 715-720, Oct. 1995.

[48] G. L. Dempsey and E. S. McVey, Circuit implementation of a peak detector neural network, *IEEE Trans. Circuits Syst. II*, vol. 40 no. 9, pp.585-591, 1993.

[49] M. Di Marco, M. Forti and A. Tesi, Existence and characterization of limit cycles in nearly symmetric neural networks, *IEEE Trans. Circuits Syst. I*, vol. 49, no. 7, pp. 979-992, 2002.

[50] K. I. Diamantaras, K. Hornik and M. G. Strintzis, Optimal linear compression under unreliable representation and robust PCA neural models, *IEEE Trans. Neural Networks*, vol. 10, no.5, pp. 1186-1195, 1999.

[51] R. Douglas, C. Koch, M. Mahowald, K. Martin and H. Suarez, Recurrent excitation in neocortical circuits, *Science*, vol. 269, pp. 981-985, 1995.

[52] P. V. D. Driessche and X. Zou, Global attractivity in delayed Hopfield neural network models, *SIAM J. Appl. Math.*, vol. 58, no. 6, pp. 1878-1890, 1998.

[53] R. Fantacci, M. Forti, M. Marini and L. Pancani, Cellular neural network approach to a class of communication problems, *IEEE Trans. Circuits Syst. I*, vol. 46, no. 12, pp. 1457-1467, 1999.

[54] J. Feng, Lyapunov functions for neural nets with non-differentiable input-output characteristics, *Neural Computation*, vol. 9, no. 1, pp. 43-49, 1997.

[55] Z. Feng and A. N. Michel, Robustness analysis of a class of discrete-time recurrent neural networks under perturbations, *IEEE Trans. Circuits and Systems I*, Vol. 46, No. 12 , pp. 482 -1486, 1999.

[56] C. Feng and R. Plamondon, On the stability analysis of delayed neural networks systems, *Neural Networks*, vol. 14, no. 9, pp. 1181-1188, 2001.

[57] M. Forti, On global asymptotic stability of a class of nonlinear systems arising in neural network theory, *J. Differential Equations*, vol. 113, pp. 246-264, 1994.

[58] M. Forti, Some extensions of a new method to analyze complete stability of neural networks, *IEEE Trans. Neural Networks*, vol. 13, no. 5, pp. 1230-1238, 2002.

[59] M. Forti and A. Tesi, A new method to analyze complete stability of PWL cellular neural networks, *Int. J. Bifurcation and Chaos*, vol. 11, no. 3, pp. 655-676, 2001.

[60] M. Forti and A. Tesi, New conditions for global stability of neural networks with application to linear and quadratic programming problems, *IEEE Trans. Circuits Syst. I*, vol. 42, no. 9, pp.354-366, 1995.

[61] M. Forti, S. Manetti and M. Marini, Necessary and sufficient conditions for absolute stability of neural networks, *IEEE Trans. Circuits Syst. I*, vol. 41, no. 7, pp.491-494, 1994.

[62] T. Fukai and S. Tanaka, A simple neural network exhibiting selective activation of neuronal ensembles: from winnner-take-all to winner-share-all, *Neural Computation*, vol. 9, pp. 77-97, 1997.

[63] Y. Fujimoto, N. Fukuda and T. Akabane, Massively parallel architectures for large scale neural network simulations, *IEEE Trans. Neural Networks*, vol. 3, no. 6, pp. 876 -888, 1992.

[64] N. Funabiki and S. Nishikawa, A binary Hopfield neural-network approach for satellite broadcast scheduling problems, *IEEE Trans. Neural Networks*, vol. 8, no. 2, pp. 441-445, 1997.

[65] A. Le Gall and V. Zissimopoulos, Extended Hopfield models for combinatorial optimization, *IEEE Trans. Neural Networks*, vol. 10, no. 1, pp. 72-80, 1999.

[66] C. L. Giles, G. M. Kuhn, and R. J. Williams. (1994). Dynamic recurrent neural networks: Theory and applications. IEEE Transactions on Neural Networks 5:153-156.

[67] M. Gilli, Stability of cellular neural networks and delayed neural networks with non-positive templates and nonmonotonic output functions, *IEEE Trans. Circuits Syst. I*, vol. 41, no. 8, pp.518-528, 1994.

[68] K. Gopalsamy and X. Z. He, Delay independent stability in bidirectional associative memory networks, *IEEE Trans. Neural Networks*, vol. 5, no. 6, pp.998-1002, 1994.

[69] K. Gopalsamy and X. Z. He, Stability in asymmetric Hopfield nets with transmission delays, *Physica D*, vol. 76, pp.344-358, 1994.

[70] K. Gopalsamy and Sariyasa, Tiem delays and stimulus-dependent pattern formation in periodic environments in isolated neurons, *IEEE Trans. Neural Networks*, vol. 13, no. 3, pp. 551-563, 2002.

[71] G. Grassi, A new approach to design cellular neural networks for associative memories, *IEEE Trans. Circuits Syst. I*, vol. 44, no. 9, pp.835-838, 1998.

[72] G. Grassi, On discrete-time cellular neural networks for associative memories, *IEEE Trans. Circuits Syst. I*, vol. 48, no. 1, pp. 107-111, 2001.

[73] Z. H. Guan and G. Chen, On delayed impulsive Hopfield neural networks,*Neural Networks*, vol. 12, no. 2, 1999.

[74] Z. H. Guan, G. Chen and Y. Qin, On equilibria, stability, and instability of Hopfield neural networks, *IEEE Trans. Neural Networks*, vol. 11, no. 2, pp. 534-540, 2000.

[75] R. L. T. Hahnloser, On the piecewise analysis of linear threshold neural networks, *Neural Networks*, vol. 11, pp. 691-697, 1998.

[76] R. L. T. Hahnloser, R. Sarpeshkar, M. A. Mahowald, R. J. Douglas, and H. S. Seung, Digital selection and analogue amplification coexit in a cortex-inspired silicon circuit, *Nature*, vol. 405, pp. 947-951, June 2000.

[77] J. K. Hale, Ordinary Differential Equations, New York, 1980.

[78] J. K. Hale, *Theory of functional differential equations*, Springer Verlag, New York, 1977.

[79] M. Hanggi, On locally regular cellular neural networks, *IEEE Trans. Circuits Syst. I*, vol. 48, no. 5, pp. 513-520.

[80] M. Hanggi and G. S. Moschytz, An analysis of CNN settling time, *IEEE Trans. Circuits Syst. I*, vol. 47, no. 1, pp. 9-24, 2000.

[81] B. He and H. Yang, A neural-network model for monotone linear asymmetric variational inequalities, *IEEE Trans. Neural Neworks*, vol. 11, no. 1, pp. 3-16, 2000.

[82] R. Hecht-Nielsen, Counter propagation networks, *Applied Optics*, vol. 26, no. 23, pp.4979-4984, 1987.

[83] M. W. Hirsch, Differential Topology, Graduate Texts in Mathematics, vol. 33, Springer-Verlag, New York, 1976.

[84] J. J. Hopfield, Nerual networks and physical systems with emergent colloective computational abilities, lit Proc. Natl. Acad. Sci. USA, vol. 79, pp. 2554-2558, April 1982.

[85] J. J. Hopfield, Neurons with graded response have collective computational properties like those of two-state neurons, *Proc. Natl. Acad. Sci. USA*, vol. 81, no. 10, pp. 3088-3092, May 1984.

[86] J. J. Hopfield and D. W. Tank, Neural computation of decisions in optimization problems, *Biol. Cybern.*, vol. 52, pp. 141-152, 1985.

[87] K. Hornik and C. M. Kuan, Convergence analysis of local feature extraction algorithms, *Neural Networks*, vol. 5, pp. 229-240, 1992.

[88] S. Hu and J. Wang, Global stability of a class of discrete-time recurrent neural networks, *IEEE Trans. Circuits Syst. I*, vol. 49, no. 8, pp. 1104–1117, Aug. 2002.

[89] S. Hui and S. H. Żak, dynamical analysis of the brain-state-in-a-box (BSB) neural models, *IEEE Trans. Neural Networks*, vol. 3, no. 1, pp. 86-94, 1992.

[90] L. Jin and M. M. Gupta, Globally asymptotic stability of discrete-time analog neural networks, *IEEE Trans. Neural Networks*, vol. 7, pp. 1024–1031, July 1996.

[91] L. Jin, P. N. Nikiforuk and M. M. Gupta, Absolute stability conditions for discrete-time recurrent neural networks, *IEEE Trans. Neural Networks*, vol. 5, no. 6, pp. 954-964, 1994.

[92] M. Joy, On the global convergence of a class of functional differential equations with applications in neural network theory, *Journal of Mathematical Analysis and Applications*, vol. 232, pp.61-81, 1999.

[93] M. Joy, Results concerning the absolute stability of delayed neural networks, *Neural Networks*, vol. 13, pp.613-616, 2000.

[94] M. P. Joy and V. Tavsanoglu, An equilibrium analysis of CNN's, *IEEE Trans. Circuits Syst. I*, vol. 45, no. 1, pp.94-98, 1998.

[95] M. P. Joy and V. Tavsanoglu, A new parameter range for the stability of opposite-sign cellular neural networks, *IEEE Trans. Circuits Syst. I*, vol. 40, no. 3, pp. 204-207, 1993.

[96] J. C. Juang, Stability analysis of Hopfield-type neural networks, *IEEE Trans. Neural Networks*, vol. 10, no. 6, pp. 1366-1374, 1999.

[97] P. Julian, R. Dogaru and L. O. Chua, A piecewise-linear simplicial coupling cell for CNN gray-level image processing, *IEEE Trans. Circuits Syst. I*, vol. 49, no. 7, pp. 904 -913, 2002.

[98] S. Kaski and T. Kohonen, Winner-take-all networks for physiological models of competitive learning, *Neural Networks*, vol. 7, no. 6, pp.973-984, 1994.

[99] E. Kaszkurewicz and A. Bhaya, On a class of globally stable neural circuits, *IEEE Trans. Circuits Syst. I*, vol. 41, no. 2, pp. 171-174, 1994.

[100] E. Kaszkurewicz and A. Bhaya, Comments on 'Necessary of sufficient conditions for absolute stability of neural networks', *IEEE Trans. Circuits Syst. I*, vol. 42, no. 8, pp. 497-499, 1995.

[101] G. I. Kechriotis and E. S. Manolakos, Hopfield neural network implementation of the optimal CDMA multiuser detector, *IEEE Trans. Neural Networks*, vol. 7, no. 1, pp. 131-141, 1996.

[102] M. P. Kennedy and L. O. Chua, Neural networks for nonlinear program-ming, *IEEE Trans. Circuits Syst.*, vol. 35, pp. 554–562, 1988.

[103] N. Kurita and K. I. Funahashi, On the Hopfield Neural Networks and Mean Field Theory. *Neural Networks*, vol. 9, no. 9, pp. 1531-1540, 1996.

[104] D. L. Lee, New stability conditions for Hopfield networks in partial simultaneous update mode, *IEEE Trans. Neural Networks*, vol. 10, no. 4, pp. 975-978, 1999.

[105] D. L. Lee, Pattern sequence recognition using a time-varying Hopfield network, *IEEE Trans. Neural Networks*, vol. 13, no. 2, pp. 330-342, 2002.

[106] C. C. Lee and J. P. de Gyvez, Color image processing in a cellular neural-network environment, *IEEE Trans. Neural Networks*, vol. 7, no. 5, pp. 1086-1098, 1996.

[107] B. W. Lee and B. J. Sheu, Modified Hopfield neural networks for retrieving the optimal solution, *IEEE Trans. Neural Networks*, vol. 2, no. 1, pp. 137-142, 1991.

[108] M. Lemmon and B.V.K.V. Kumar, Competitive learning with generalized winner-take-all activation, *IEEE Trans. Neural Networks*, vol. 3, no. 2, pp.167-175, 1992.

[109] W. J. Li and T. Lee, Hopfield neural networks for affine invariant matching, *IEEE Trans. Neural Networks*, vol. 12, no. 6, pp. 1400-1410, 2001.

[110] X. B. Liang, Equivalence between local exponential stability of the unique equilibrium point and global stability for Hopfield-type neural networks with two neurons, *IEEE Trans. Neural Networks*, vol. 11, no. 5, pp. 1194-1196, 2000.

[111] X. B. Liang and J. Si, Global exponential stability of neural networks with globallly Lipschitz continuous activations and its applications to linear variational inequality problem, *IEEE Trans. Neural Networks*, vol. 12, no. 2, pp. 349-359, 2001.

[112] X. B. Liang and S. K. Tso, An improved upper bound on step-size parameters of discrete-time recurrent neural networks for linear inequality and equation system, *IEEE Trans. Circuits Syst. I*, vol. 49, no. 5, pp. 695-698, 2002.

[113] X. B. Liang and J. Wang, Absolute exponential stability of neural networks with a general class of activation functions, *IEEE Trans. Circuits and Systems, I*, vol. 47, no. 8, pp. 1258-1263, 2000.

[114] X. B. Liang and J. Wang, A recurrent neural networks for nonlinear optimization with a continuously differentiable objective function and bounded constranints, *IEEE Trans. Neural Networks*, vol. 11, no. 6, pp. 1251-1262, 2000.

[115] X. B. Liang and L. D. Wu, Comments on 'New conditions for global stability of neural networks with applications to linear and quadratic programming problems', *IEEE Trans. Circuits Syst. I*, vol. 44, no. 11, pp.1099-1101, 1997.

[116] X. B. Liang and L. D. Wu, New sufficient conditions for absolute stability of neural networks, *IEEE Trans. Circuits Syst. I*, vol. 45, no. 4, pp. 584-586, 1998.

[117] X. F. Liao, G. Chen and E. N. Sanchez, Delay-dependent exponential stability analysis of delayed neural networks: an LMI approach, *Nerual Networks*, vol. 15, no. 7, pp. 855-866, 2002.

[118] X. Liao and J. Yu, Robust stability for interval Hopfield neural networks with time delay, *IEEE Trans. Neural Networks*, vol. 9, no. 5, pp.1042-1045, 1998.

[119] T. L. Liao and F. C. Wang, Global stability for cellular neural networks with time delay, *IEEE Trans. Neural Networks*, vol. 11, no. 6, pp. 1481-1484, 2000.

[120] X. F. Liao, K. W. Wong and J. B. Yu, Novel stability conditions for cellular neural networks with time delay, *Int. J. Bifurcation and Chaos*, vol. 11, no. 7, pp. 1853-1864, 2001.

[121] X. X. Liao, Mathematic theory of cellular neural networks .2., *Science in China, Ser A*, vol. 38, no. 5, pp. 542-551, 1995.

[122] X. X. Liao, Stability of Hopfield-type neural neural networks .1., *Science in China, ser. A*, vol. 38, no. 4, pp. 407-418, 1995.

[123] X. X. Liao and Y. Liao, Stability of Hopfield-type neural networks .2., *Science in China, Ser. A*, vol. 40, no. 8, pp. 813-816, 1997.

[124] S. S. Lin and C. W. Shih, Complete stability for standard cellular neural networks, *Int. J. Bifurcation and Chaos*, vol. 9, pp. 909-918, 1999.

[125] X. Liu and T. Chen, A new result on the global convergence of Hopfield neural networks, *IEEE Trans. Circuits Syst. I*, vol. 49, no. 10, pp. 1514-1516, 20002.

[126] D. Liu and A. N. Michel, Asymptotic stability of discrete-time systems with saturation nonlinearities with applications to digital filters, *IEEE Trans. Circuits Syst. I*, vol. 39, pp. 798–807, Oct. 1992.

[127] H. Lu, On stability of nonlinear continuous-time neural networks with delays, *Neural Networks*, vol. 13, pp. 1135-1143, 2000.

[128] H. Lu, Stability criteria for delayed neural networks, *Physical Review E*, 64, 051901, 2001.

[129] H. Lu, Y. He and Z. He, A chaos-generator: analyses of complex dynamics of a cell equation in delayed cellular neural networks, *IEEE Trans. Circuits Syst. I*, vol. 45, no. 2, pp. 178-181, 1998.

[130] F. Luo, R. Unbehauen and A. Cichocki, A minor component analysis algorithm, *Neural Networks*, vol. 10, no. 2, PP. 291-297, 1997.

[131] F. Luo and R. Unbehauen, A minor subspace analysis algorithm, *IEEE Trans. Neural Networks*, vol. 8, no, 5, pp. 1149-1153.

[132] F. Luo, R. Unbehauen and Y. D. Li, A principal component analysis algorithm with invariant norm, *Neurocomputing*, vol. 8, pp. 213-221, 1995.

[133] J. Ma, Simplex memory neural networks, *Neural Networks*, vol. 10, no. 1, pp. 25-29, 1997.

[134] J. Ma, The Stability of the Generalized Hopfield Networks in Randomly Asynchronous Mode, vol. 10, no. 6, pp. 1109-1116, 1997.

[135] J. Ma, The asymptotic memory capacity of the generalized Hopfield network, vol. 12, no. 9, pp. 1207-1212, 1999.

[136] C. Y. Maa and M. Shanblatt, Linear and quadratic programming neural network analysis, *IEEE Trans. Neural Networks*, vol. 3, no. 4, pp. 580–594, 1992.

[137] C. M. Marcus and R. M. Westervelt, Stability of analog neural networks with delay, *Physical Review A*, Vol. 39, no. 1, pp.347-359, 1989.

[138] G. Mathew and V. U. Reddy, Development and analysis of a neural network approach to Pisarenko's harmonic retrieval method, *IEEE Trans. Signal Processing*, vol. 42, no.3, pp.663-667, 1994.

[139] G. Mathew and V. U. Reddy, Orthogonal eigensubspace estimation using neural networks, *IEEE Trans. Signal Processing*, vol. 42, no.7, pp.1803-1811, 1994.

[140] G. Mathew, V. U. Reddy and S. Dasgupta, Adaptive estimation of eigensubspace, *IEEE Trans. Signal Processing*, vol. 43, no. 2, pp. 401-411, 1995.

[141] J. E. McInroy and B. M. Wilamowski, Bipolar pattern association using a recurrent winner take all network, *Proceedings of 1997 International Conference on Neural Networks (ICNN'97)*, vol. 2, pp.1231-1234, 1997.

[142] A. N. Michel, J. Si and G. Yen, Analysis and synthesis of a class of discrete-time neural networks described on hypercubes, *IEEE Trans. Neural Networks*, vol. 2, no. 1, pp. 32-46, 1991.

[143] S. Mitra, S. K. Pal and P. Mitra, Data mining in soft computing framework: a survey, *IEEE Trans. Neural Networks*, vol. 13, no. 1, pp. 3-14, 2002.

[144] , V. M. Mladenov, D. M. W. Leenaerts and F. H. Uhlmann, Estimation of the basin of attractions in CNNs, *IEEE Trans. Circuits Syst. I*, vol. 45, no. 5, pp. 571-574, 1998.

[145] N. M. Nasrabadi and C. Y. Choo, Hopfield network for stereo vision correspondence, *IEEE Trans. Neural Networks*, vol. 3, no. 1, pp. 5-13, 1992.

[146] L. Nemes and T. Roska, A CNN model of oscillation and chaos in ant colonies: a case study, *IEEE Trans. Circuits Syst. I*, vol. 42, no. 10, pp. 741-745, 1995.

[147] E. Oja and J. Karhunen, On stochastic approximation of the eigenvectors and eigenvalues of the expectation of a random matrix, *J. of Math. Anal. Appl.*, vol. 106, pp.69-84, 1985.

[148] E. Oja, Principal components, minor components, and linear neural networks, *Neural Networks* , vol. 5, pp. 927-935, 1992.

[149] K. Pakdaman, C. G. Ragazzo and C. P. Malta, Transient reigme duration in continuous-time neuaral networks with delay, *Physical Review E*, vol. 58, no. 3, pp. 3623-3627, 1998.

[150] D. C. Park and Y. J. Woo, Weighted centroid neural network for edge preserving image compression, *IEEE Trans. Neural Networks*, vol. 12, no. 5, 2001.

[151] M. Peng, N. K. Gupta and A. F. Armitage, An investigation into the improvement of local minima of the Hopfield network, *Neural Networks*, vol. 9, no. 7, pp. 1241-1253, 1996.

[152] J. Peng, H. Qiao and Z. B. Xu, A new approach to stability of neural networks with time-varying delays, vol. 15, no. 1, pp. 95-103, 2002.

[153] M. J. Pérez-Ilzarbe, Convergence analysis of a discrete-time recurrent neural networks to perform quadratic real optimization with bound constraints, *IEEE Trans. Neural Networks*, vol. 9, no. 6, pp. 1344-1351, 1998.

[154] R. Perfetti, A synthesis procedure for brain-state-in-a-box neural networks, *IEEE Trans. Neural Networks*, vol. 6, pp. 1071-1080, 1995.

[155] R. Perfetti, Sensitivity of equilibrium points in continuous-time Hopfield's network, *IEEE Trans. Neural Network*, vol. 6, no. 2, pp. 501-504, 1995.

[156] H. Qiao, J. Peng and Z. B. Xu, Nonlinear measures: a new approach to exponential stability analysis for Hopfield-type neural networks, *IEEE Trans. Neural Networks*, vol. 12 no. 2, pp: 360-370, 2001.

[157] K. Reif, F. Lou and R. Unbehauen, The exponential stability of the invariant norm PCA algorithm, *IEEE Trans. Circuits and Sys.-II*, vol. 44, no. 10, pp.873-876, 1997.

[158] T. Roska and L. O. Chua, Cellular neural networks with delay type template elements and nonuniform grids, *Int. J. Circuit Theory and applicat.*, vol. 20, pp.469-481, 1992.

[159] T. Roska, C. W. Wu, M. Balsi and L. O. Chua, Stability and dynamics of delay-type general and cellular neural networks, *IEEE Trans. Circuits Syst. I*, vol. 39, no. 6, pp.487-490, 1992.

[160] T. Roska, C. W. Wu and L. O. Chua, Stability of cellular neural networks with dominant nonlinear and delay-type templates, *IEEE Trans. Circuits Syst. I*, vol. 40, no. 4, pp. 270-272, 1993.

[161] E. N. Sanchez and J. P. Perez, Input-to-State stability (ISS) anlysis for dynamic neural networks, *IEEE Trans. Circuits and Systems, I*, vol. 46, no. 11, pp. 1395-1398, 1999.

[162] G. Seiler and J. Nossek, Winner-take-all cellular neural networks, *IEEE Trans. Circuits Syst. II*, vol. 40, no. 3, pp.184-190, 1993.

[163] H. S. Seung, How the brain keeps the eyes still, *Proc. Natl. Acad. Sci. USA*, vol. 93, pp. 13339-13344, 1996.

[164] G. Setti, P. Thiran and C. Serpico, An approach to information propagation in 1-D cellular neural networks. II. Global propagation, *IEEE Trans. Circuits Syst. I*, vol. 45, no. 8, pp. 790-811, 1998.

[165] B. E. Shi and L. O. Chua, Resistive grid image filtering: input/output analysis via the CNN framework, *IEEE Trans. Circuits Syst. I*, vol. 39, no. 7, pp. 531-548, 1992.

[166] J. Si and A. N. Michel, Analysis and synthesis of a class of discrete-time neural networks with multilevel threshold neurons, *IEEE Trans. Neual Networks*, vol. 6, no. 1, pp. 105-116, 1995.

[167] A. Slavova, Dynamic properties of cellular neural networks with nonlinear output function, *IEEE Trans. Circuits Syst. I*, vol. 45, no. 5, pp. 587-590, 1998.

[168] E. D. Sontag, Further facts about input to state stabilization, *IEEE Trans. Automatic Control*, vol. 35, no. 4, pp. 473-476, 1990.

[169] E. D. Sontag, Smooth stabilization implies coprime factorization, *IEEE Trans. Automatic Control*, vol. 34, pp. 435-443, 1989.

[170] E. D. Sontag and Y. Wang, On characterizations of the input-to-state stability property, *Systems & Control Letters*, vol. 24, pp. 351-359, 1995.

[171] E. D. Sontag and Y. Wang, New characterizations of input-to-state stability, *IEEE Trans. Automatic Control*, vol. 41, no. 9, pp. 1283-1294, 1996.

[172] A. Srinivasan and C. Batur, Hopfield/ART-1 neural network-based fault detection and isolation, *IEEE Trans. Neural Networks*, vol. 5, no. 6, pp. 890-899, 1994.

[173] V. Sree Hari Rao and Bh. R. M. Phaneendra, Global dynamics of bidirectional associative memory neural networks involving transmission delays and dead zones, vol. 12, no. 3, pp. 455-465, 1999.

[174] S. I. Sudharsanan and M. K. Sundareshan, Exponential stability and a systematic synthesis of a neural network for quadratic minimization, *Neural Networks*, vol. 4, pp. 599–613, 1991.

[175] I. Szatmari, A. Schultz, C. Rekeczky, T. Kozek, T. Roska, and L. O. Chua, Morphology and autowave metric on CNN applied to bubble-debris classification, *IEEE Trans. Neural Networks*, vol. 11, no. 6, pp. 1385-1393, 2000.

[176] T. Sziranyi and J. Zerubia, Markov random field image segmentation using cellular neural network, *IEEE Trans. Circuits Syst. I*, vol. 44, no. 1, pp. 86-89, 1997.

[177] C. Sun, A comment on "Global stability analysis in delayed Hopfield neural network models", *Neural Networks*, vol. 15, no. 10, pp. 1299-1300, 2002.

[178] J. P. F. Sum, C. S. Leung, P. K. S. Tam, G. H. Young, W. K. Kan and L. W. Chan, Analysis for a class of winner-take-all model, *IEEE Trans. Neural Networks*, vol. 10, no. 1, pp.64-71, 1999.

[179] N. Takahashi, A new sufficient condition for complete stability of cellular neural networks with delay, *IEEE Trans. Circuits Syst. I*, vol. 47, no. 6, pp.793-799, 2000.

[180] N. Takahashi and L. O. Chua, A new sufficient condition for nonsymmetric CNNs to have a stable equilibrium point, *IEEE Trans. Circuits Syst. I*, vol. 44, no. 11, pp. 1092-1095, 1997.

[181] N. Takahashi and L. O. Chua, On the complete stability of nonsymmetric cellular neural networks, *IEEE Trans. Circuits Syst. I*, vol. 45, pp. 754-758, July, 1998.

[182] P. M. Talavan and J. Yanez, Parameter setting of the Hopfield network applied to TSP, *Neural Neworks*, vol. 15, no. 3, pp. 363-373, 2002.

[183] A. Taleb and G. Cirrincione, Against the convergence of the minor component analysis neurons, *IEEE Trans. Neural Networks*, vol. 10, no. 1, pp.207-210, 1999.

[184] D. W. Tank and J. J. Hopfield, Simple neural optimization networks: An A/D converter, signal decision circuit, and a linear programming circuit, *IEEE Trans. Circuits Systems*, vol. 33, pp. 533-541, 1986.

[185] P. Thiran, Influence of boundary conditions on the behavior of cellular neural networks, *IEEE Trans. Circuits Syst. I*, vol. 40, no. 3, pp. 207-212, 1993.

[186] P. Thiran, G. Setti and M. Hasler, An approach to information propagation in 1-D cellular neural networks-Part I: Local diffusion, *IEEE Trans. Circuits Syst. I*, vol. 45, no. 8, pp. 77-789, 1998.

[187] A. Torralba, F. Colodro, E. Ibanez and L. G. Freanquelo, Two digital circuits for a fully parallel stochastic neural network, *IEEE Trans. Neural Networks*, vol. 6, no. 5, pp. 1264-1268, 1995.

[188] S. Townley, A. Ilchmann, M. G. Wei$\beta$, W. Mcclements, A.C. Ruiz, D. H. Owens and D. Prätzel-Wolters, Existence and learning of oscillations in recurrent neural networks, *IEEE Trans. Neural Networks*, vol. 11, no. 1, pp. 205-214, 2000.

[189] K. Urahama and T. Nagao, K-Winner-take-all circuit with O(N) complexity, *IEEE Trans. Neural Networks*, vol. 6, no. 3, pp.776-778, 1995.

[190] D. E. Van Den Bout and T. K. Miller, Graph partitioning using annealed neural networks, *IEEE Trans. Neural Networks*, vol. 1, no. 2, pp. 192-203, 1990.

[191] P. L. Venetianer, T. Roska, Image compression by delayed CNN's, *IEEE Trans. Circuits Syst. I*, vol. 45, no. 3, pp.205-215, 1998.

[192] J. Wang, A recurrent neural network for solving shortest path problem, *IEEE Trans. Circuits Systems, I*, vol. 43, no. 6, pp. 482-486, 1996.

[193] J. Wang, Analysis and design of a recurrent neural networks for linear programming, *IEEE Trans. Circuits and Systems, I*, vol. 40, no. 9, pp. 613-618, 1993.

[194] J. Wang, Analysis and design of an analog sorting network, *IEEE Trans. Neural Networks*, vol. 6, no. 4, pp. 962-971, 1995.

[195] J. Wang, Analogue neural networks for solving the assignment problem, *Electronics Letters*, vol. 28, no. 11, pp. 1047-1050, 1992.

[196] J. Wang, Primal and dual assignment networks, *IEEE Trans. Neural Networks*, vol. 8, no. 3, pp. 784-790, 1997.

[197] L. Wang, Multi-associative neural networks and their applications to learning and retrieving complex spatio-temporal sequences, *IEEE Trans. Systems, Man, and Cybernetics - Part B: Cybernetics*, vol. 29, no. 1, pp.73-82, Feb 1999.

[198] L. Wang, On competitive learning, *IEEE Trans. Neural Networks*, vol. 8, no. 5, pp.1214-1217, Sept 1997.

[199] J. S. Wang, Q. Gan and Y. Wei, Stability of CNN with opposite-sign templates and nonunity gain output functions, *IEEE Trans. Circuits Syst. I*, vol. 42, no. 7, pp. 404-408, 1995.

[200] K. Wang and A. N. Michel, On the stability of a family of nonlinear time-varying systems, *IEEE Trans. Circuits Syst. I*, vol. 43, no. 7, pp. 517-531, 1996.

[201] L. Wang and K. Smith, On chaotic simulated annealing, *IEEE Trans. Neural Networks*, vol. 9, no. 4, pp. 716-718, 1998.

[202] T. Washizawa, Application of Hopfield network to saccades, *IEEE Trans. Neural Networks*, vol. 4, no. 6, pp. 995-997, 1993.

[203] H. Wersing, W. J. Beyn and H. Ritter, Dynamical stability conditions for recurrent neural networks with unsaturating piecewise linear transfer functions, *Neural Computation*, vol. 13, pp. 1811-1825, 2001.

[204] H. Wersing, J. J. Steil and H. Ritter, A competitive layer model for feature binding and sensory segmentation, *Neural Computation*, vol. 13, no. 2, PP. 357-387, 2001.

[205] C. W. Wu, and L. O. Chua, A more rigorous preof of complete stability of cellular neural networks, *IEEE Trans. Circuits and Systems, I*, vol. 44, no. 4, pp. 370-371, 1997.

[206] Y. Xia and J. Wang, A general methodology for designing globally convergent optimization neural networks, *IEEE Trans. Neural Networks*, vol. 9, pp. 1331–1343, 1998.

[207] Y. Xia, J. Wang, and D. L. Hung, Recurrent neural networks for solving linear inequalities and equations, *IEEE Trans. Circuits Syst. I*, vol. 46, pp. 452–462, Apr. 1999.

[208] X. Xie, R. H. R. Hahnloser and H. S. Seung, Selective grouping neurons in recurrent networks of lateral inhibition, *Neural Computation*, vol. 14, no. 11, pp. 2627-2646, 2002.

[209] Z. B. Xu, G. Q. Hu and C. P. Kwong, Asymmetric Hopfield-type Networks: Theory and Applications, *Neural Networks*, vol. 9, no. 3, pp. 483-501, 1996.

[210] L. Xu, E. Oja and C. Suen, Modified Hebbian learning for curve and surface fitting, *Neural Networks*, vol. 5, pp.441-457, 1992.

[211] H. Ye, A. N. Michel and K. Wang, Global stability and local stability of Hopfield neural networks with delays, *Physical Review E*, vol. 50, no. 5, pp. 4206-4213, 1994.

[212] J. C. Yen, J. I. Guo, and H.-C. Chen, A new k -winners-take all neural network and its array architecture, *IEEE Trans. Neural Networks*, vol. 9, pp. 901–912, Sept. 1998.

[213] J. F. Yang and C.M. Chen, Winner-take-all neural networks using the highest threshold. *IEEE Trans. Neural Networks*, Vol. 11 No. 1, pp. 194 -199, 2000.

[214] J. F. Yang and M. Kaveh, Adaptive eigensubspace algorithms for direction or frequency estimation and tracking, *IEEE Trans. Acoust., Speech, Signal Processing*, vol. 36, no. 2, pp.663-667, 1988.

[215] H. Yang and T. S. Dillon, Exponential stability and oscillation of Hopfield graded response neural network, *IEEE Trans. Neural Networks*, vol. 5, no. 5, pp. 719-729, 1994.

[216] T. Yang and L. B. Yang, Application of fuzzy cellular neural networks to Euclidean distance transformation, *IEEE Trans. Circuits and Systems, I*, vol. 44, no. 3, pp. 242-246, 1997.

[217] Z. Yi, P. A. Heng and P. F. Fung, Winner-take-all discrete recurrent neural networks, *IEEE Trans. Circuits and Systems, II*, vol. 47, no. 12, pp. 1584-1589, Feb. 2000.

[218] Z. Yi, P. A. Heng and K. S. Leung, Convergence analysis of cellular neural networks with unbounded delay, *IEEE Trans. Circuits and Systems, I*, vol. 48, no. 6, pp. 680-687, 2001.

[219] Z. Yi, P. A. Heng and P. Vadakkepat, Absolute periodicity and absolute stability of delayed neural networks, *IEEE Trans. Circuits and Systems, I*, vol. 49, no. 2, 2002.

[220] Z. Yi, P. A. Heng and A. W. C. Fu, Estimate of exponential convergence rate and exponential stability for neural networks, *IEEE Trans. Neural Networks*, vol. 10, no. 6, pp. 1487-1493, 1999.

[221] Z. Yi and P. A. Heng, Stability of fuzzy control systems with bounded uncertain delays, *IEEE Trans. Fuzzy Systems*, vol. 10, no. 1, pp. 92-97, 2002.

[222] Z. Yi and K. K. Tan, Dynamic stability conditions for Lotka-Volterra recurrent neural networks with delays, *Phys. Rev. E*, vol. 66, 011910, 2002.

[223] Z. Yi, K. K. Tan and T. H. Lee, Multistability analysis for recurrent neural networks with ujnsaturating piecewise linear transfer functions, *Nerual Computation*, vol. 15, no. 3, pp. 639-662, 2003.

[224] Z. Yi, Global exponential stability and periodic solutions of delay Hopfield neural networks, *Int. J. Syst. Sci.*, vol. 27, no. 2, pp.227-231, 1996.

[225] Z. Yi, S. M. Zhong and Z. L. Li, Periodic solutions and stability of Hopfield neural networks with variable delays, *Int. J. Syst. Sci.*, vol. 27, no. 9, pp.895-901, 1996.

[226] Z. Yi, J. B. Yu and Y. Wu, Global stability analysis on a class of cellular neural networks, *Science in China, Ser. A*, vol. 44, no. 1, pp. 1-11, 2001.

[227] A. Yuille and N. Grzywacz, A winner-take-all mechanism based on presynaptic inhibition feedback, *Neural Comput.*, vol. 1, pp.334-347, 1989.

[228] A. Zarandy, L. Orzo, E. Grawes and F. Werblin, CNN-based models for color vision and visual illusions, *IEEE Trans. Circuits Syst. I*, vol. 46, no. 2, pp. 229-238, 1999.

[229] A. Zarandy, A. Stoffels, A. Roska and L. O. Chua, Implementation of binary and gray-scale mathematical morphology on the CNN universal machine, *IEEE Trans. Circuits Syst. I*, vol. 45, no. 2, pp. 163-168, 1988.

[230] K. Zhang, Representation of spatial orientation by the intrinsic dynamics of the head-direction cell ensemble: a theory, *J. Neurosci.*, vol.16, pp. 2112-2126, 1996.

[231] J. Zhang and X. Jin, Global stability analysis in delayed Hopfield neural networks models, *Neural Networks*, 13, pp. 745-753, 2000.

GPSR Compliance

*The European Union's (EU) General Product Safety Regulation (GPSR) is a set of rules that requires consumer products to be safe and our obligations to ensure this.*

*If you have any concerns about our products, you can contact us on ProductSafety@springernature.com*

In case Publisher is established outside the EU, the EU authorized representative is:

Springer Nature Customer Service Center GmbH
Europaplatz 3
69115 Heidelberg, Germany

Batch number: 10607214

Printed by Printforce, the Netherlands